DANA G. MARSH

QUALITY ENGINEERING
USING
ROBUST DESIGN

QUALITY ENGINEERING
USING
ROBUST DESIGN

MADHAV S. PHADKE

AT&T Bell Laboratories

Prentice Hall PTR, Englewood Cliffs, New Jersey 07632

Library of Congress Cataloging-in-Publication Data

Phadke, Madhav Shridhar
    Quality engineering using robust design / Madhav S. Phadke.
        p.   ca.
    Includes index.
    ISBN 0-13-745167-9
    1. Engineering design. 2. Computer-aided design. 3. UNIX
(Computer operating system) 4. Integrated circuits--Very large
scale integration. I. Title.
TA174.P49 1989
620 ' .0042 ' 0285--dc20   89-3927
                                                           CIP

© 1989 by AT&T Bell Laboratories

Published by P T R Prentice-Hall, Inc.
A Simon & Schuster Company
Englewood Cliffs, New Jersey 07632

Printed in the United States of America

20 19 18 17 16 15 14 13 12

ISBN 0-13-745167-9

ISBN 0-13-745167-9

90000

9 780137 451678

Prentice-Hall International (UK) Limited, *London*
Prentice-Hall of Australia Pty. Limited, *Sydney*
Prentice-Hall Canada Inc., *Toronto*
Prentice-Hall Hispanoamericana, S.A., *Mexico*
Prentice-Hall of India Private Limited, *New Delhi*
Prentice-Hall of Hapan, Inc., *Tokyo*
Simon & Schuster Asia Pte. Ltd., *Singapore*
Editora Prentice-Hall do Brasil, Ltda., *Rio de Janeiro*

To my parents, and Maneesha, Kedar, and Lata.

CONTENTS

FOREWORD

The main task of a design engineer is to build in the function specified by the product planning people at a competitive cost. An engineer knows that all kinds of functions are energy transformations. Therefore, the product designer must identify what is input, what is output, and what is ideal function while developing a new product. It is important to make the product's function as close to the ideal function as possible.

Therefore, it is very important to measure correctly the distance of the product's performance from the ideal function. This is the main role of quality engineering. In order to measure the distance, we have to consider the following problems:

1. Identify signal and noise space

2. Select several points from the space

3. Select an adequate design parameter to observe the performance

4. Consider possible calibration or adjustment method

5. Select an appropriate measurement related with the mean distance

As most of those problems require engineering knowledge, a book on quality engineering must be written by a person who has enough knowledge of engineering.

Dr. Madhav Phadke, a mechanical engineer, has worked at AT&T Bell Laboratories for many years and has extensive experience in applying the Robust Design method to problems from diverse engineering fields. He has made many eminent and pioneering contributions in quality engineering, and he is one of the best qualified persons to author a book on quality engineering.

The greatest strength of this book is the case studies. Dr. Phadke presents four real instances where the Robust Design method was used to improve the quality and cost of products. Robust Design is universally applicable to all engineering fields. You will be able to use these case studies to improve the quality and cost of your products.

This is the first book on quality engineering, written in English by an engineer. The method described here has been applied successfully in many companies in Japan, USA, and other countries. I recommend this book for all engineers who want to apply experimental design for actual product design.

G. Taguchi

PREFACE

Designing high-quality products and processes at low cost is an economic and technological challenge to the engineer. A systematic and efficient way to meet this challenge is a new method of design optimization for performance, quality, and cost. The method, called Robust Design, consists of

1. Making product performance insensitive to raw material variation, thus allowing the use of low grade material and components in most cases,

2. Making designs robust against manufacturing variation, thus reducing labor and material cost for rework and scrap,

3. Making the designs least sensitive to the variation in operating environment, thus improving reliability and reducing operating cost, and

4. Using a new structured development process so that engineering time is used most productively.

All engineering designs involve setting values of a large number of decision variables. Technical experience together with experiments, through prototype hardware models or computer simulations, are needed to come up with the most advantageous decisions about these variables. Studying these variables one at a time or by trial and error is the common approach to the decision process. This leads to either a very long and expensive time span for completing the design or premature termination of the design process so that the product design is nonoptimal. This can mean missing the market window and/or delivering an inferior quality product at an inflated cost.

The Robust Design method uses a mathematical tool called *orthogonal arrays* to study a large number of decision variables with a small number of experiments. It also uses a new measure of quality, called *signal-to-noise* (S/N) *ratio*, to predict the quality from the customer's perspective. Thus, the most economical product and process design from both manufacturing and customers' viewpoints can be accomplished at the smallest, affordable development cost. Many companies, big and small, high-tech and low-tech, have found the Robust Design method valuable in making high-quality products available to customers at a low competitive price while still maintaining an acceptable profit margin.

This book will be useful to practicing engineers and engineering managers from all disciplines. It can also be used as a text in a quality engineering course for seniors and first year graduate students. The method is explained through a series of real case studies, thus making it easy for the readers to follow the method without the burden of learning detailed theory. At AT&T, several colleagues and I have developed a two and a half day course on this topic. My experience in teaching the course ten times has convinced me that the case studies approach is the best one to communicate how to use the method in practice. The particular case studies used in this book relate to the fabrication of integrated circuits, circuit design, computer tuning, and mechanical routing.

Although the book is written primarily for engineers, it can also be used by statisticians to study the wide range of applications of experimental design in quality engineering. This book differs from the available books on statistical experimental design in that it focuses on the engineering problems rather than on the statistical theory. Only those statistical ideas that are relevant for solving the broad class of product and process design problems are discussed in the book.

Chapters 1 through 7 describe the necessary theoretical and practical aspects of the Robust Design method. The remaining chapters show a variety of applications from different engineering disciplines. The best way for readers to use this book is, after reading each section, to determine how the concepts apply to their projects. My experience in teaching the method has revealed that many engineers like to see an application of the method in their own field. Chapters 8 through 11 describe case studies from different engineering fields. It is hoped that these case studies will help readers see the breadth of the applicability of the Robust Design method and assist them in their own applications.

Madhav S. Phadke

AT&T Bell Laboratories
Holmdel, N.J.

ACKNOWLEDGMENTS

I had the greatest fortune to learn the Robust Design methodology directly from its founder, Professor Genichi Taguchi. It is with the deepest gratitude that I acknowledge his inspiring work. My involvement in the Robust Design method began when Dr. Roshan Chaddha asked me to host Professor Taguchi's visit to AT&T Bell Laboratories in 1980. I thank Dr. Chaddha (Bellcore, formerly with AT&T Bell Labs) for the invaluable encouragement he gave me during the early applications of the method in AT&T and also while writing this book. I also received valuable support and encouragement from Dr. E. W. Hinds, Dr. A. B. Godfrey, Dr. R. E. Kerwin, and Mr. E. Fuchs in applying the Robust Design method to many different engineering fields which led to deeper understanding and enhancement of the method.

Writing a book of this type needs a large amount of time. I am indebted to Ms. Cathy Savolaine for funding the project. I also thank Mr. J. V. Bodycomb and Mr. Larry Bernstein for supporting the project.

The case studies used in this book were conducted through collaboration with many colleagues, Mr. Gary Blaine, Mr. Dave Chrisman, Mr. Joe Leanza, Dr. T. W. Pao, Mr. C. S. Sherrerd, Dr. Peter Hey, and Mr. Paul Sherry. I am grateful to them for allowing me to use the case studies in the book.

I also thank my colleagues, Mr. Don Speeney, Dr. Raghu Kackar, and Dr. Mike Grieco, who worked with me on the first Robust Design case study at AT&T. Through this case study, which resulted in huge improvements in the window photolithography process used in integrated circuits fabrication, I gained much insight into the Robust Design method.

I thank Mr. Rajiv Keny for numerous discussions on the organization of the book. A number of my colleagues read the draft of the book and provided me with valuable comments. Some of the people who provided the comments are: Dr. Don Clausing (M.I.T.), Dr. A. M. Joglekar (Honeywell), Dr. C. W. Hoover, Jr. (Polytechnic University), Dr. Jim Pennell (IDA), Dr. Steve Eick, Mr. Don Speeney, Dr. M. Daneshmand, Dr. V. N. Nair, Dr. Mike Luvalle, Dr. Ajit S. Manocha, Dr. V. V. S. Rana, Ms. Cathy Hudson, Dr. Miguel Perez, Mr. Chris Sherrerd, Dr. M. H. Sherif, Dr. Helen Hwang, Dr. Vasant Prabhu, Ms. Valerie Partridge, Dr. Sachio Nakamura, Dr. K. Dehnad, and Dr. Gary Ulrich. I thank them all for their generous help in improving the content and readability of the book. I also thank Mr. Akira Tomishima (Yamatake-Honeywell), Dr. Mohammed Hamami, and Mr. Bruce Linick for helpful discussions on specific topics in the book. Thanks are also due to Mr. Yuin Wu (ASI) for valuable general discussions.

I very much appreciate the editorial help I received from Mr. Robert Wright and Ms. April Cormaci through the various stages of manuscript preparation. Also, I thank Ms. Eve Engel for coordinating text processing and the artwork during manuscript preparation.

The text of this volume was prepared using the UNIX* operating system, 5.2.6a, and a LINOTRONIC® 300 was used to typeset the manuscript. Mr. Wright was responsible for designing the book format and coordinating production. Mr. Don Hankinson, Ms. Mari-Lynn Hankinson, and Ms. Marilyn Tomaino produced the final illustrations and were responsible for the layout. Ms. Kathleen Attwooll, Ms. Sharon Morgan, and several members of the Holmdel Text Processing Center provided electronic text processing.

Chapter 1

INTRODUCTION

The objective of engineering design, a major part of research and development (R&D), is to produce drawings, specifications, and other relevant information needed to manufacture products that meet customer requirements. Knowledge of scientific phenomena and past engineering experience with similar product designs and manufacturing processes form the basis of the engineering design activity (see Figure 1.1). However, a number of new decisions related to the particular product must be made regarding product architecture, parameters of the product design, the process architecture, and parameters of the manufacturing process. A large amount of engineering effort is consumed in conducting experiments (either with hardware or by simulation) to generate the information needed to guide these decisions. Efficiency in generating such information is the key to meeting market windows, keeping development and manufacturing costs low, and having high-quality products. *Robust Design* is an engineering methodology for improving productivity during research and development so that high-quality products can be produced quickly and at low cost.

This chapter gives an overview of the basic concepts underlying the Robust Design methodology:

- Section 1.1 gives a brief historical background of the method.

- Section 1.2 defines the term *quality* as it is used in this book.

- Section 1.3 enumerates the basic elements of the cost of a product.

- Section 1.4 describes the fundamental principle of the Robust Design methodology with the help of a manufacturing example.

- Section 1.5 briefly describes the major tools used in Robust Design.

- Section 1.6 presents some representative problems and the benefits of using the Robust Design method in addressing them.

- Section 1.7 gives a chapter-by-chapter outline of the rest of the book.

- Section 1.8 summarizes the important points of this chapter.

In the subsequent chapters, we describe Robust Design concepts in detail and, through case studies, we show how to apply them.

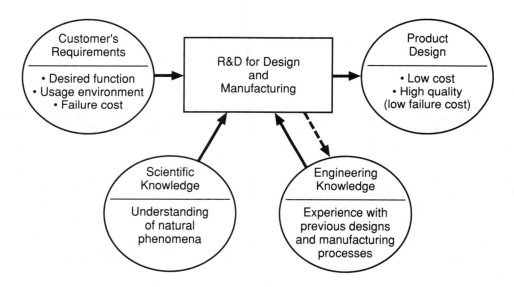

Figure 1.1   Block diagram of R&D activity.

1.1 A HISTORICAL PERSPECTIVE

When Japan began its reconstruction efforts after World War II, it faced an acute short-age of good-quality raw material, high-quality manufacturing equipment and skilled engineers. The challenge was to produce high-quality products and continue to improve the quality under those circumstances. The task of developing a methodology to meet the challenge was assigned to Dr. Genichi Taguchi, who at that time was a manager in charge of developing certain telecommunications products at the Electrical Communications Laboratories (ECL) of Nippon Telephone and Telegraph Company (NTT). Through his research in the 1950s and the early 1960s, Dr. Taguchi developed the foundations of Robust Design and validated its basic philosophies by applying them in the development of many products. In recognition of this contribution, Dr. Taguchi received the individual Deming Award in 1962, which is one of the highest recognitions in the quality field.

The Robust Design method can be applied to a wide variety of problems. The application of the method in electronics, automotive products, photography and many other industries has been an important factor in the rapid industrial growth and the subsequent domination of international markets in these industries by Japan.

Robust Design draws on many ideas from statistical experimental design to plan experiments for obtaining dependable information about variables involved in making engineering decisions. The science of statistical experimental design originated with the work of Sir Ronald Fisher in England in the 1920s. Fisher founded the basic principles of experimental design and the associated data-analysis technique called *analysis of variance* (ANOVA) during his efforts to improve the yield of agricultural crops. The theory and applications of experimental design and the related technique of *response surface methodology* have been advanced by many statistical researchers. Today, many excellent textbooks on this subject exist, for example, Box, Hunter and Hunter [B3], Box and Draper [B2], Hicks [H2], John [J2], Raghavarao [R1], and Kempthorne [K4]. Various types of matrices are used for planning experiments to study several decision variables simultaneously. Among them, Robust Design makes heavy use of the *orthogonal arrays*, whose use for planning experiments was first proposed by Rao [R2].

Robust Design adds a new dimension to statistical experimental design—it explicitly addresses the following concerns faced by all product and process designers:

- How to reduce economically the variation of a product's function in the customer's environment. (Note that achieving a product's function consistently on target maximizes customer satisfaction.)

- How to ensure that decisions found to be optimum during laboratory experiments will prove to be so in manufacturing and in customer environments.

In addressing these concerns, Robust Design uses the mathematical formalism of statistical experimental design, but the thought process behind the mathematics is different in many ways. The answers provided by Robust Design to the two concerns listed above make it a valuable tool for improving the productivity of the R&D activity.

The Robust Design method is still evolving. With the active research being carried out in the United States, Japan, and other countries, it is expected that the applications of the method and the method itself will grow rapidly in the coming decade.

1.2 WHAT IS QUALITY?

Because the word *quality* means different things to different people (see, for example, Juran [J3], Deming [D2], Crosby [C5], Garvin [G1], and Feigenbaum [F1]), we need to define its use in this book. First, let us define what we mean by *the ideal quality* which can serve as a reference point for measuring the *quality level* of a product. The ideal quality a customer can expect is that every product delivers the target performance each time the product is used, under all intended operating conditions, and throughout its intended life, with no harmful side effects. Note that the traditional concepts of reliability and dependability are part of this definition of quality. In specific situations, it may be impossible to produce a product with ideal quality. Nonetheless, ideal quality serves as a useful reference point for measuring the quality level.

The following example helps clarify the definition of ideal quality. People buy automobiles for different purposes. Some people buy them to impress their friends while others buy them to show off their social status. To satisfy these diverse purposes, there are different types (species) of cars—sports cars, luxury cars, etc.—on the market. For any type of car, the buyer always wants the automobile to provide reliable transportation. Thus, for each type of car, an ideal quality automobile is one that works perfectly each time it is used (on hot summer days and cold winter days), throughout its intended life (not just the warranty life) and does not pollute the atmosphere.

When a product's performance deviates from the target performance, its quality is considered inferior. The performance may differ from one unit to another or from one environmental condition to another, or it might deteriorate before the expiration of the intended life of the product. Such deviation in performance causes loss to the user of the product, the manufacturer of the product, and, in varying degrees, to the rest of the society as well. Following Taguchi, *we measure the quality of a product in terms of the total loss to society due to functional variation and harmful side effects.* Under the ideal quality, the loss would be zero; the greater the loss, the lower the quality.

In the automobile example, if a car breaks down on the road, the driver would, at the least, be delayed in reaching his or her destination. The disabled car might be the cause of traffic jams or accidents. The driver might have to spend money to have the car towed. If the car were under warranty, the manufacturer would have to pay for repairs. The concept of quality loss includes all these costs, not just the warranty cost. Quantifying the quality loss is difficult and is discussed in Chapter 2.

Note that the definition of quality of a product can be easily extended to processes as well as services. As a matter of fact, the entire discussion of the Robust Design method in this book is equally applicable for processes and services, though for simplicity, we do not state so each time.

1.3 ELEMENTS OF COST

Quality at what cost? Delivering a high-quality product at low cost is an interdisciplinary problem involving engineering, economics, statistics, and management. The three main categories of cost one must consider in delivering a product are:

1. *Operating Cost.* Operating cost consists of the cost of energy needed to operate the product, environmental control, maintenance, inventory of spare parts and units, etc. Products made by different manufacturers can have different energy costs. If a product is sensitive to temperature and humidity, then elaborate and costly air conditioning and heating units are needed. A high product failure rate of a product causes large maintenance costs and costly inventory of spare units. A manufacturer can greatly reduce the operating cost by designing the product robust—that is, minimizing the product's sensitivity to environmental and usage conditions, manufacturing variation, and deterioration of parts.

2. *Manufacturing Cost.* Important elements of manufacturing cost are equipment, machinery, raw materials, labor, scrap, rework, etc. In a competitive environment, it is important to keep the *unit manufacturing cost* (umc) low by using low-grade material, employing less-skilled workers, and using less-expensive equipment, and at the same time maintain an appropriate level of quality. This is possible by designing the product robust, and designing the manufacturing process robust—that is, minimizing the process' sensitivity to manufacturing disturbances.

3. *R&D Cost.* The time taken to develop a new product plus the amount of engineering and laboratory resources needed are the major elements of R&D cost. The goal of R&D activity is to keep the umc and operating cost low. Robust Design plays an important role in achieving this goal because it improves the efficiency of generating information needed to design products and processes, thus reducing development time and resources needed for development.

Note that the manufacturing cost and R&D cost are incurred by the producer and then passed on to the customer through the purchase price of the product. The operating cost, which is also called *usage cost*, is borne directly by the customer and it is directly related to the product's quality. From the customer's point of view, the purchase price plus the operating cost determine the economics of satisfying the need for which the product is bought. Higher quality means lower operating cost and vice versa. Robust Design is a systematic method for keeping the producer's cost low while delivering a high-quality product, that is, while keeping the operating cost low.

1.4 FUNDAMENTAL PRINCIPLE

The key idea behind Robust Design is illustrated by the experience of Ina Tile Company, described in detail in Taguchi and Wu [T7]. During the late 1950s, Ina Tile Company in Japan faced the problem of high variability in the dimensions of the tiles it produced [see Figure 1.2(a)]. Because *screening* (rejecting those tiles outside specified dimensions) was an expensive solution, the company assigned a team of expert engineers to investigate the cause of the problem. The team's analysis showed that the tiles at the center of the pile inside the kiln [see Figure 1.2 (b)] experienced lower temperature than those on the periphery. This nonuniformity of temperature distribution proved to be the cause of the nonuniform tile dimensions. The team reported that it would cost approximately half a million dollars to redesign and build a kiln in which all the tiles would receive uniform temperature distribution. Although this alternative was less expensive than screening it was still too costly.

The team then brainstormed and defined a number of process parameters that could be changed easily and inexpensively. After performing a small set of well-planned experiments according to Robust Design methodology, the team concluded that increasing the lime content of the clay from 1 percent to 5 percent would greatly reduce the variation of the tile dimensions. Because lime was the least expensive ingredient, the cost implication of this change was also favorable.

Thus, the problem of nonuniform tile dimensions was solved by minimizing the effect of the cause of the variation (nonuniform temperature distribution) without controlling the cause itself (the kiln design). As illustrated by this example, the fundamental principle of Robust Design is to improve the quality of a product by minimizing the effect of the causes of variation without eliminating the causes. This is achieved by optimizing the product and process designs to make the performance minimally sensitive to the various causes of variation. This is called *parameter design*. However, parameter design alone does not always lead to sufficiently high quality. Further improvement can be obtained by controlling the causes of variation where economically justifiable, typically by using more expensive equipment, higher grade components, better environmental controls, etc., all of which lead to higher product cost, or operating cost, or both. The benefits of improved quality must justify the added product cost.

1.5 TOOLS USED IN ROBUST DESIGN

A great deal of engineering time is spent generating information about how different design parameters affect performance under different usage conditions. Robust Design methodology serves as an "amplifier"—that is, it enables an engineer to generate information needed for decision-making with half (or even less) the experimental effort.

There are two important tasks to be performed in Robust Design:

1. *Measurement of Quality During Design/Development.* We want a leading indicator of quality by which we can evaluate the effect of changing a particular design parameter on the product's performance.

2. *Efficient Experimentation to Find Dependable Information about the Design Parameters.* It is critical to obtain dependable information about the design parameters so that design changes during manufacturing and customer use can be avoided. Also, the information should be obtained with minimum time and resources.

The estimated effects of design parameters must be valid even when other parameters are changed during the subsequent design effort or when designs of related subsystems change. This can be achieved by employing the *signal-to-noise (S/N) ratio* to measure quality and *orthogonal arrays* to study many design parameters simultaneously. These tools are described later in this book.

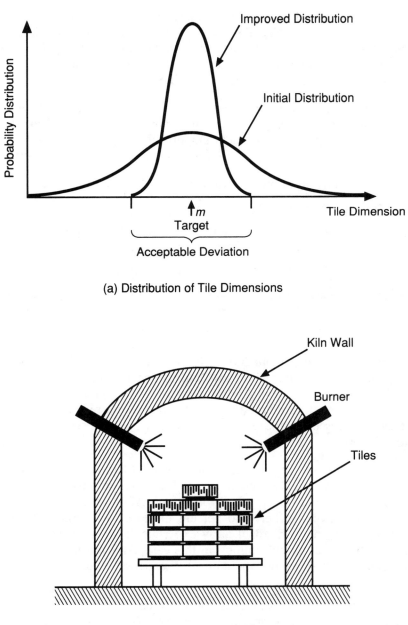

(a) Distribution of Tile Dimensions

(b) Schematic Diagram of the Kiln

Figure 1.2   Tile manufacturing example.

1.6 APPLICATIONS AND BENEFITS OF ROBUST DESIGN

The Robust Design method is in use in many areas of engineering throughout the United States. For example, AT&T's use of Robust Design methodology has lead to improvement of several processes in *very large scale integrated (VLSI) circuit fabrication* used in the manufacture of 1-megabit and 256-kilobit memory chips, 32-bit processor chips, and other products. Some of the VLSI applications are:

- The *window photolithography* application (documented in Phadke, Kackar, Speeney, and Grieco [P5]) was the first application in the United States that demonstrated the power of Taguchi's approach to quality and cost improvement through robust process design. In particular, the benefits of the application were:

    — 4-fold reduction in process variance

    — 3-fold reduction in fatal defects

    — 2-fold reduction in processing time (because the process became stable, allowing time-consuming inspection to be dropped)

    — Easy transition of design from research to manufacturing

    — Easy adaptation of the process to finer-line technology (adaptation from 3.5-micron to 2.5-micron technology), which is typically a very difficult problem.

- The *aluminum etching* application originated from a belief that poor photoresist print quality leads to line width loss and to undercutting during the etching process. By making the etching process insensitive to photoresist profile variation and other sources of variation, the visual defects were reduced from 80 percent to 15 percent. Moreover, the etching step could then tolerate the variation in the photoresist profile.

- The *reactive ion etching* of tantalum silicide (described in Katz and Phadke [K3]), used to give highly nonuniform etch quality, so only 12 out of 18 possible wafer positions could be used for production. After optimization, 17 wafer positions became usable—a hefty 40 percent increase in machine utilization. Also, the efficiency of the orthogonal array experimentation allowed this project to be completed by the 20-day deadline. In this case, $1.2 million was saved in equipment replacement costs not including the expense of disruption on the factory floor.

- The *polysilicon deposition* process had between 10 and 5000 surface defects per unit area. As such, it represented a serious road block in advancing to line widths smaller than 1.75 micron. Six process parameters were investigated with 18 experiments leading to consistently less than 10 surface defects per unit area. As a result, the scrap rate was reduced significantly and it became possible to process smaller line widths. This case study is described in detail in Chapter 4.

Other AT&T applications include:

* The *router bit life-improvement* project (described in Chapter 11 and Phadke [P3]) led to a 2-fold to 4-fold increase in the life of router bits used in cutting printed wiring boards. The project illustrates how reliability or life improvement projects can be organized to find the best settings of the routing process parameters with a very small number of samples. The number of samples needed in this approach is very small, yet it can give valuable information about how each parameter changes the survival probability curve (change in survival probability as a function of time).

* In the *differential operational amplifier circuit optimization* application (described in Chapter 8 and Phadke [P3]), a 40-percent reduction in the root mean square (rms) offset voltage was realized by simply finding new nominal values for the circuit parameters. This was done by reducing sensitivity to all tolerances and temperature, rather than reducing tolerances, which could have increased manufacturing cost.

* The Robust Design method was also used to find optimum proportions of ingredients for making *water-soluble flux*. By simultaneous study of the parameters for the wave soldering process and the flux composition, the defect rate was reduced by 30 to 40 percent (see Lin and Kackar [L3]).

* Orthogonal array experiments can be used to *tune hardware/software systems*. By simultaneous study of three hardware and six software parameters, the response time of the UNIX operating system was reduced 60 percent for a particular set of load conditions experienced by the machine (see Chapter 10 and Pao, Phadke, and Sherrerd [P1]).

Under the leadership of American Supplier Institute and Ford Motor Company, a number of automotive suppliers have achieved quality and cost improvement through Robust Design. These applications include improvements in metal casting, injection molding of plastic parts, wave soldering of electronic components, speedometer cable design, integrated circuit chip bonding, and picture tube lens coating. Many of these applications are documented in the Proceedings of Supplier Symposia on Taguchi Methods [P9].

All these examples show that the Robust Design methodology offers simultaneous improvement of product quality, performance and cost, and engineering productivity. Its widespread use in industry will have a far-reaching economic impact because this methodology can be applied profitably in all engineering activities, including product design and manufacturing process design.

The philosophy behind Robust Design is not limited to engineering applications. Yokoyama and Taguchi [Y1] have also shown its applications in profit planning in business, cash-flow optimization in banking, government policymaking, and other areas. The method can also be used for tasks such as determining optimum work force mix for jobs where the demand is random, and improving the runway utilization at an airport.

1.7 ORGANIZATION OF THE BOOK

This book is divided into three parts. The first part (Chapters 1 through 4) describes the basics of the Robust Design methodology. Chapter 2 describes the quality loss function, which gives a quantitative way of evaluating the quality level of a product rather than just the "good-bad" characterization. After categorizing the sources of variation, the chapter further describes the steps in engineering design and the classification of parameters affecting the product's function. Quality control activities during different stages of the product realization process are also described there. Chapter 3 is devoted to orthogonal array experiments and basic analysis of the data obtained through such experiments. Chapter 4 illustrates the entire strategy of Robust Design through an integrated circuit (IC) process design example. The strategy begins with problem formulation and ends with verification experiment and implementation. This case study could be used as a model in planning and carrying out manufacturing process optimization for quality, cost, and manufacturability. The example also has the basic framework for optimizing a product design.

The second part of the book (Chapters 5 through 7) describes, in detail, the techniques used in Robust Design. Chapter 5 describes the concept of signal-to-noise ratio and gives appropriate signal-to-noise ratios for a number of common engineering problems. Chapter 6 is devoted to a critical decision in Robust Design: choosing an appropriate response variable, called *quality characteristic*, for measuring the quality of a product or a process. The guidelines for choosing quality characteristics are illustrated with examples from many different engineering fields. A step-by-step procedure for designing orthogonal array experiments for a large variety of industrial problems is given in Chapter 7.

The third part of the book (Chapters 8 through 11) describes four more case studies to illustrate the use of Robust Design in a wide variety of engineering disciplines. Chapter 8 shows how the Robust Design method can be used to optimize product design when computer simulation models are available. The differential operational amplifier case study is used to illustrate the optimization procedure. This chapter also shows the use of orthogonal arrays to simulate the variation in component values and environmental conditions, and thus estimate the yield of a product. Chapter 9 shows the procedure for designing an ON-OFF control system for a temperature controller. The use of Robust Design for improving the performance of a hardware-software system is described in Chapter 10 with the help of the UNIX operating system tuning case study. Chapter 11 describes the router bit life study and explains how Robust Design can be used to improve reliability.

1.8 SUMMARY

- Robust Design is an engineering methodology for improving productivity during research and development so that high-quality products can be produced quickly and at low cost. Its use can greatly improve an organization's ability to meet market windows, keep development and manufacturing costs low, and deliver high-quality products.

- Through his research in the 1950s and early 1960s, Dr. Genichi Taguchi developed the foundations of Robust Design and validated the basic, underlying philosophies by applying them in the development of many products.

- Robust Design uses many ideas from statistical experimental design and adds a new dimension to it by explicitly addressing two major concerns faced by all product and process designers:

  a. How to reduce economically the variation of a product's function in the customer's environment.

  b. How to ensure that decisions found optimum during laboratory experiments will prove to be so in manufacturing and in customer environments.

- The *ideal quality* a customer can receive is that every product delivers the target performance each time the product is used, under all intended operating conditions, and throughout the product's intended life, with no harmful side effects. The deviation of a product's performance from the target causes loss to the user of the product, the manufacturer, and, in varying degrees, to the rest of society as well. The *quality level* of a product is measured in terms of the total loss to the society due to functional variation and harmful side effects.

- The three main categories of cost one must consider in delivering a product are: (1) *operating cost*: the cost of energy, environmental control, maintenance, inventory of spare parts, etc. (2) *manufacturing cost:* the cost of equipment, machinery, raw materials, labor, scrap, network, etc. (3) *R&D cost*: the time taken to develop a new product plus the engineering and laboratory resources needed.

- The fundamental principle of Robust Design is to improve the quality of a product by minimizing the effect of the causes of variation without eliminating the causes. This is achieved by optimizing the product and process designs to make the performance minimally sensitive to the various causes of variation, a process called *parameter design*.

- The two major tools used in Robust Design are: (1) *signal-to-noise ratio*, which measures quality and (2) *orthogonal arrays*, which are used to study many design parameters simultaneously.

- The Robust Design method has been found valuable in virtually all engineering fields and business applications.

Chapter 2

PRINCIPLES OF
QUALITY ENGINEERING

A product's life cycle can be divided into two main parts: before sale to the customer and after sale to the customer. All costs incurred prior to the sale of the product are added to the unit manufacturing cost (umc), while all costs incurred after the sale are lumped together as *quality loss*. *Quality engineering* is concerned with reducing both of these costs and, thus, is an interdisciplinary science involving engineering design, manufacturing operations, and economics.

It is often said that higher quality (lower quality loss) implies higher unit manufacturing cost. Where does this misconception come from? It arises because engineers and managers, unaware of the Robust Design method, tend to achieve higher quality by using more costly parts, components, and manufacturing processes. In this chapter we delineate the basic principles of quality engineering and put in perspective the role of Robust Design in reducing the quality loss as well as the umc. This chapter contains nine sections:

- Sections 2.1 and 2.2 are concerned with the quantification of quality loss. Section 2.1 describes the shortcomings of using *fraction defective* as a measure of quality loss. (This is the most commonly used measure of quality loss.) Section 2.2 describes the *quadratic loss function*, which is a superior way of quantifying quality loss in most situations.

- Section 2.3 describes the various causes, called *noise factors*, that lead to the deviation of a product's function from its target.

- Section 2.4 focuses on the computation of the *average quality loss,* its components, and the relationship of these components to the noise factors.

- Section 2.5 describes how Robust Design exploits nonlinearity to reduce the average quality loss without increasing umc.

- Section 2.6 describes the classification of parameters, an important activity in quality engineering for recognizing the different roles played by the various parameters that affect a product's performance.

- Section 2.7 discusses different ways of formulating product and process design optimization problems and gives a heuristic solution.

- Section 2.8 addresses the various stages of the *product realization process* and the role of various quality control activities in these stages.

- Section 2.9 summarizes the important points of this chapter.

Various aspects of quality engineering are described in the following references: Taguchi [T2], Taguchi and Wu [T7], Phadke [P2], Taguchi and Phadke [T6], Kackar [K1,K2], Taguchi [T4], Clausing [C1], and Byrne and Taguchi [B4].

2.1 QUALITY LOSS FUNCTION—THE FRACTION DEFECTIVE FALLACY

We have defined the quality level of a product to be the total loss incurred by society due to the failure of the product to deliver the target performance and due to harmful side effects of the product, including its operating cost. Quantifying this loss is difficult because the same product may be used by different customers, for different applications, under different environmental conditions, etc. However, it is important to quantify the loss so that the impact of alternative product designs and manufacturing processes on customers can be evaluated and appropriate engineering decisions made. Moreover, it is critical that the quantification of loss not become a major task that consumes substantial resources at various stages of product and process design.

It is common to measure quality in terms of the fraction of the total number of units that are defective. This is referred to as *fraction defective.* Although commonly used, this measure of quality is often incomplete and misleading. It implies that all products that meet the specifications (allowable deviations from the target response) are equally good, while those outside the specifications are bad. The fallacy here is that the product that barely meets the specifications is, from the customer's point of view, as good or as bad as the product that is barely outside the specifications. In reality, the product whose response is exactly on target gives the best performance. As the product's response deviates from the target, the quality becomes progressively worse.

Example—Television Set Color Density:

The deficiency of fraction defective as a quality measure is well-illustrated by the Sony television customer preference study published by the Japanese newspaper, *The Asahi* [T8]. In the late 1970s, American consumers showed a preference for the television sets made by Sony-Japan over those made by Sony-USA. The reason cited in the study was quality. Both factories, however, made televisions using identical designs and tolerances. What could then account for the perceived difference in quality?

In its investigative report, the newspaper showed the distribution of color density for the sets made by the two factories (see Figure 2.1). In the figure, m is the target color density and $m \pm 5$ are the *tolerance limits* (allowable manufacturing deviations). The distribution for the Sony-Japan factory was approximately normal with mean on target and a standard deviation of 5/3. The distribution for Sony-USA was approximately uniform in the range of $m \pm 5$. Among the sets shipped by Sony-Japan, about 0.3 percent were outside the tolerance limits, while Sony-USA shipped virtually no sets outside the tolerance limits. Thus, the difference in customer preference could not be explained in terms of the fraction defective sets.

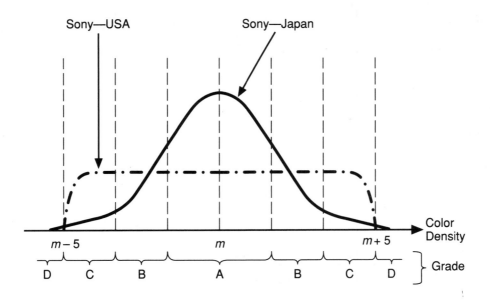

Figure 2.1   Distribution of color density in television sets. (Source: *The Asahi*, April 17, 1979).

The perceived difference in quality becomes clear when we look closely at the sets that met the tolerance limits. Sets with color density very near m perform best and can be classified *grade* A. As the color density deviates from m, the performance becomes progressively worse, as indicated in Figure 2.1 by *grades* B and C. It is clear that Sony-Japan produced many more grade A sets and many fewer grade C sets when compared to Sony-USA. Thus, the average grade of sets produced by Sony-Japan was better, hence the customer's preference for the sets made by Sony-Japan.

In short, the difference in the customer's perception of quality was a result of Sony-USA paying attention only to *meeting the tolerances,* whereas in Sony-Japan the attention was focused on *meeting the target.*

Example—Telephone Cable Resistance:

Using a wrong measurement system can, and often does, drive the behavior of people in wrong directions. The telephone cable example described here illustrates how using fraction defective as a measure of quality loss can permit suboptimization by the manufacturer leading to an increase in the *total cost*, which is the sum of quality loss and umc.

A certain gauge of copper wires used in telephone cables had a nominal resistance value of m ohms/mile and the maximum allowed resistance was $(m + \Delta_0)$ ohms/mile. This upper limit was determined by taking into consideration the manufacturing capability, represented by the distribution (a) in Figure 2.2, at the time the specifications were written. Consequently, the upper limit $(m + \Delta_0)$ was an adequate way to ensure that the drawing process used to form the copper wire was kept in control with the mean on target.

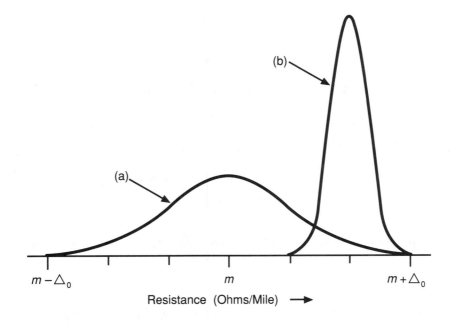

Figure 2.2   Distribution of telephone cable resistance. (a) Initial distribution. (b) After process improvement and shifting the mean.

By improving the wire drawing process through the application of new technology, the manufacturer was able to reduce substantially the process variance. This permitted the manufacturer to move the mean close to the upper limit and still meet the fraction defective criterion for quality [see distribution (b) in Figure 2.2]. At the same time, the manufacturer saved on the cost of copper since larger resistance implies a smaller cross section of the wire. However, from the network point of view, the larger average resistance resulted in high electrical loss, causing complaints from the telephone users. Solving the problem in the field meant spending a lot more money for installing additional repeaters and for other corrective actions than the money saved in manufacturing—that is, the increase in the quality loss far exceeded the saving in the umc. Thus, there was a net loss to the society consisting of both the manufacturer and the telephone company who offered the service. Therefore, a quality loss metric that permits such local optimization leading to higher total cost should be avoided. Section 2.2 discusses a better way to measure the quality loss.

Interpretation of Engineering Tolerances

The examples above bring out an important point regarding quantification of quality loss. Products that do not meet tolerances inflict a quality loss on the manufacturer, a loss visible in the form of scrap or rework in the factory, which the manufacturer adds to the cost of the product. However, *products that meet tolerance also inflict a quality loss, a loss that is visible to the customer and that can adversely affect the sales of the product and the reputation of the manufacturer.* Therefore, the quality loss function must also be capable of measuring the loss due to products that meet the tolerances.

Engineering specifications are invariably written as $m \pm \Delta_0$. These specifications should *not* be interpreted to mean that any value in the range $(m - \Delta_0)$ to $(m + \Delta_0)$ is equally good for the customer and that as soon as the range is exceeded the product is bad. In other words, the step function shown below and in Figure 2.3(a) is an inadequate way to quantify the quality loss:

$$L(y) = \begin{cases} 0 & \text{if } |y-m| \le \Delta_0 \\ A_0 & \text{otherwise} \end{cases} \tag{2.1}$$

Here, A_0 is the cost of replacement or repair. Use of such a loss function is apt to lead to the problems that Sony-USA and the cable manufacturer faced and, hence, should be avoided.

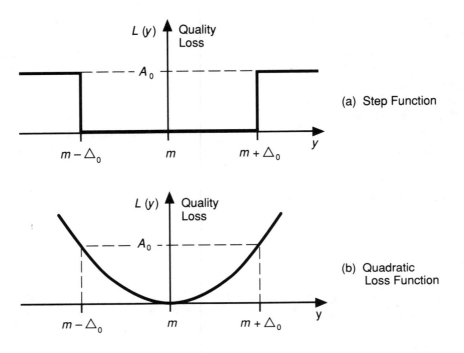

Figure 2.3   Quality loss function.

2.2 QUADRATIC LOSS FUNCTION

The quadratic loss function can meaningfully approximate the quality loss in most situations. Let y be the quality characteristic of a product and m be the target value for y. (Note that the quality characteristic is a product's response that is observed for quantifying quality level and for optimization in a Robust Design project.) According to the quadratic loss function, the quality loss is given by

$$L(y) = k(y-m)^2 \qquad (2.2)$$

where k is a constant called *quality loss coefficient*. Equation (2.2) is plotted in Figure 2.3(b). Notice that at $y = m$ the loss is zero and so is the slope of the loss function. This is quite appropriate because m is the best value for y. The loss $L(y)$ increases slowly when we are near m; but as we go farther from m the loss increases more rapidly. Qualitatively, this is exactly the kind of behavior we would like the quality loss function to have. The quadratic loss function given by Equation (2.2) is the simplest mathematical function that has the desired qualitative behavior.

Note that Equation (2.2) does not imply that every customer who receives a product with y as the value of the quality characteristic will incur a precise quality loss equal to $L(y)$. Rather, it implies that the average quality loss incurred by those customers is $L(y)$. The quality loss incurred by a particular customer will obviously depend on that customer's operating environment.

It is important to determine the constant k so that Equation (2.2) can best approximate the actual loss within the region of interest. This is a rather difficult, though important, task. A convenient way to determine k is to determine first the *functional limits* for the value of y. Functional limit is the value of y at which the product would fail in half of the applications. Let $m \pm \Delta_0$ be the functional limits. Suppose, the loss at $m \pm \Delta_0$ is A_0. Then by substitution in Equation (2.2), we obtain

$$k = \frac{A_0}{\Delta_0^2} \ . \tag{2.3}$$

Note that A_0 is the cost of repair or replacement of the product. It includes the loss due to the unavailability of the product during the repair period, the cost of transporting the product by the customer to and from the repair center, etc. If a product fails in an unsafe mode, such as an automobile breaking down in the middle of a road, then the losses from the resulting consequences should also be included in A_0. Regardless of who pays for them—the customer, the manufacturer, or a third party—all these losses should be included in A_0.

Substituting Equation (2.3) in Equation (2.2) we obtain

$$L\ (y) = \frac{A_0}{\Delta_0^2}(y - m)^2 \ . \tag{2.4}$$

We will now consider two numerical examples.

Example—Television Set Color Density:

Suppose the functional limits for the color density are $m \pm 7$. This means about half the customers, taking into account the diversity of their environment and taste, would find the television set to be defective if the color density is $m \pm 7$. Let the repair of a television set in the field cost on average $A_0 = \$98$. By substituting in Equation (2.4), the quadratic loss function can be written as

$$L(y) = \frac{98}{7^2}\ (y - m)^2 = 2(y - m)^2 \ .$$

Thus, the average quality loss incurred by the customers receiving sets with color density $m + 4$ is $L(m + 4) = \$32$, while customers receiving sets with color density $m + 2$ incur an average quality loss of only $L(m + 2) = \$8$.

Example—Power Supply Circuit:

Consider a power supply circuit used in a stereo system for which the target output voltage is 110 volts. If the output voltage falls outside 110 ± 20 volts, then the stereo fails in half the situations and must be repaired. Suppose it costs \$100 to repair the stereo. Then the average loss associated with a particular value y of output voltage is given by

$$L(y) = \frac{100}{20^2} (y - 110)^2 = 0.25(y - 110)^2 \quad .$$

Variations of the Quadratic Loss Function

The quadratic loss function given by Equation (2.2) is applicable whenever the quality characteristic y has a finite target value, usually nonzero, and the quality loss is symmetric on either side of the target. Such quality characteristics are called *nominal-the-best* type quality characteristics and Equation (2.2) is called the nominal-the-best type quality loss function. The color density of a television set and the output voltage of a power supply circuit are examples of the nominal-the-best type quality characteristic.

Some variations of the quadratic loss function in Equation (2.2) are needed to cover adequately certain commonly occurring situations. Three such variations are given below.

- *Smaller-the-better type characteristic.* Some characteristics, such as radiation leakage from a microwave oven, can never take negative values. Also, their ideal value is equal to zero, and as their value increases, the performance becomes progressively worse. Such characteristics are called *smaller-the-better* type quality characteristics. The response time of a computer, leakage current in electronic circuits, and pollution from an automobile are additional examples of this type of quality characteristic. The quality loss in such situations can be approximated by the following function, which is obtained from Equation (2.2) by substituting $m = 0$:

$$L(y) = ky^2 \quad . \tag{2.5}$$

  Note this is a one-sided loss function because y cannot take negative values. As described earlier, the quality loss coefficient k can be determined from the functional limit, Δ_0, and the quality loss, A_0, can be determined at the functional limit by using Equation (2.3).

- *Larger-the-better type characteristic.* Some characteristics, such as the bond strength of adhesives, also do not take negative values. But, zero is their worst value, and as their value becomes larger, the performance becomes progressively better—that is, the quality loss becomes progressively smaller. Their ideal value is infinity and at that point the loss is zero. Such characteristics are called *larger-the-better* type quality characteristics. It is clear that the reciprocal of such a characteristic has the same qualitative behavior as a smaller-the-better type characteristic. Thus, we approximate the loss function for a larger-the-better type characteristic by substituting $1/y$ for y in Equation (2.5):

$$L(y) = k \left[\frac{1}{y^2} \right] . \tag{2.6}$$

The rationale for using Equation (2.6) as the quality loss function for larger-the-better type characteristics is discussed further in Chapter 5. To determine the constant k for this case, we find the functional limit, Δ_0, below which more than half of the products fail, and the corresponding loss A_0. Substituting Δ_0 and A_0 in Equation (2.6), and solving for k, we obtain

$$k = A_0 \Delta_0^2 . \tag{2.7}$$

- *Asymmetric loss function.* In certain situations, deviation of the quality characteristic in one direction is much more harmful than in the other direction. In such cases, one can use a different coefficient k for the two directions. Thus, the quality loss would be approximated by the following asymmetric loss function:

$$L(y) = \begin{cases} k_1(y-m)^2, & y > m \\ k_2(y-m)^2, & y \le m . \end{cases} \tag{2.8}$$

The four different versions of the quadratic loss function are plotted in Figure 2.4. For a more detailed discussion of the quality loss function see Taguchi [T4] and Jessup [J1].

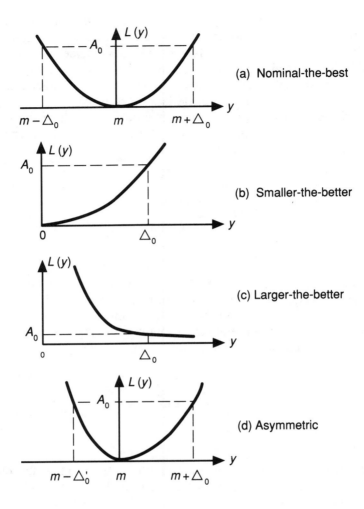

Figure 2.4   Variations of the quadratic loss function.

Quantifying Quality Loss Internal to a Company

The notion of quality loss should not be applied with regard to only the external customer of a company. In fact, it can and should be applied at every step of product realization within a company. For example, some 200 steps are required in manufacturing integrated circuits. The efficiency and costs at an intermediate step are affected by the variations introduced in the preceding steps; the intermediate step, in turn, affects the efficiency and costs of the subsequent steps. The quadratic loss function provides a reasonable way to evaluate these impacts on cost and efficiency and, thus, facilitates appropriate economic decisions regarding quality improvement at various process steps.

2.3 NOISE FACTORS—CAUSES OF VARIATION

The performance of a product as measured by the quality characteristic varies in the field due to a variety of causes. We call all such causes *noise factors*, and they can be generally classified as follows:

1. *External.* The environment in which a product works and the load to which it is subjected are the two main external sources of variation of a product's function. Some examples of environmental noise factors are temperature, humidity, dust, supply voltage, electromagnetic interference, vibrations, and human error in operating the product. The number of tasks to which a product is subjected simultaneously and the period of time a product is exercised continuously are two examples of load-related noise factors.

2. *Unit-to-unit variation.* The variation that is inevitable in a manufacturing process leads to variation in the product parameters from unit to unit. For example, the value of a resistor may be specified to be 100 kilo-ohms, but the resistance value turns out to be 101 kilo-ohms in one particular unit and 98 kilo-ohms in another.

3. *Deterioration.* When a product is sold, all its functional characteristics may be on target. As time passes, however, the values of individual components may change leading to deterioration in product performance.

Let us look at a few common products and identify key noise factors:

- *Refrigerator.* Some of the important noise factors related to the temperature control inside a refrigerator are:

    — *External*—the number of times the door is opened and closed, the amount of food kept and the initial temperature of the food, variation in the ambient temperature, and the supply voltage variation.

    — *Unit-to-unit variation*—the tightness of the door closure and the amount of refrigerant used.

    — *Deterioration*—the leakage of refrigerant and mechanical wear of compressor parts.

- *Automobile.* The following noise factors are important for the braking distance of an automobile:

    — *External*—wet or dry road, concrete or asphalt pavement, and number of passengers in the car.

    — *Unit-to-unit variation*—variation in the friction coefficient of the pads and drums, and amount of brake fluid.

    — *Deterioration*—the wear of the drums and brake pads, and leakage of brake fluid.

Manufacturing processes, which are the source of unit-to-unit variation, are themselves affected by many sources of variation. We also call these sources *noise factors.* These noise factors can be classified into three categories similar to those described in reference to a product's performance. The three categories are:

1. *External to the process.* These are the noise factors related to the environment in which the process is carried out (ambient temperature, humidity, etc.) and the load offered to the process. Variation in the raw material and operator errors are also examples of this category.

2. *Process nonuniformity.* In some processes, many units are processed simultaneously as a batch. For example, in wave soldering of printed circuit boards as many as 1000 or more solder joints may be formed simultaneously. Each solder joint experiences different processing conditions based on its position on the board. In some processes, process nonuniformity can be an important source of variation.

3. *Process drift.* Due to the depletion of chemicals used or the wearing out of the tools, the average quality characteristic of the products may drift as more units are produced.

The following example further clarifies the three types of noise factors in a manufacturing process:

*   *Developing photos.* Some of the key noise factors for the developing process are:

    — *External to the process*—the number of films being developed simultaneously and lighting conditions in the room.

    — *Process nonuniformity*—the variation in the develop time from batch to batch, and variation in the reactivity of the chemicals at different points inside the processing machine.

    — *Process drift*—change in the reactivity of the chemicals used as more batches of film are developed.

As discussed in Chapter 1, the fundamental principle of Robust Design is to improve the quality by minimizing the effect of the causes of variation. Thus, it is important in every Robust Design project to identify important noise factors. For products or processes that have multiple functions, different noise factors can affect different quality characteristics. For example, in xerographic machines, noise factors for image quality could be different from those for paper transport. In such situations, the different quality characteristics must be optimized separately. Engineering experience and judgment are needed in identifying the noise factors. Efficient experimentation for sorting out more important noise factors from less important noise factors is discussed in Chapter 8.

2.4 AVERAGE QUALITY LOSS

Because of the noise factors, the quality characteristic y of a product varies from unit to unit, and from time to time during the usage of the product. Suppose the distribution of y resulting from all sources of noise is as shown in Figure 2.5. Let $y_1, y_2, \cdots, y_n$ be n representative measurements of the quality characteristic y taken on a few representative units throughout the design life of the product. Let y be a nominal-the-best type quality characteristic and m be its target value. Then, the average quality loss, Q, resulting from this product is given by

$$
\begin{aligned}
Q &= \frac{1}{n} \left[L(y_1) + L(y_2) + \cdots + L(y_n) \right] \\[2mm]
&= \frac{k}{n} \left[(y_1 - m)^2 + (y_2 - m)^2 + \cdots + (y_n - m)^2 \right] \\[2mm]
&= k \left[(\mu - m)^2 + \frac{n-1}{n}\, \sigma^2 \right]
\end{aligned}
\tag{2.9}
$$

where μ and σ^2 are the mean and the variance of y, respectively, as computed by

$$
\mu = \frac{1}{n} \sum_{i=1}^{n} y_i \quad \text{and} \quad \sigma^2 = \frac{1}{n-1} \sum_{i=1}^{n} (y_i - \mu)^2 \quad .
$$

When n is large, Equation (2.9) can be written as

$$
Q = k \left[(\mu - m)^2 + \sigma^2 \right] .
\tag{2.10}
$$

Thus, the average quality loss has the following two components:

1. $k(\mu - m)^2$ resulting from the deviation of the average value of y from the target

2. $k\sigma^2$ resulting from the mean squared deviation of y around its own mean

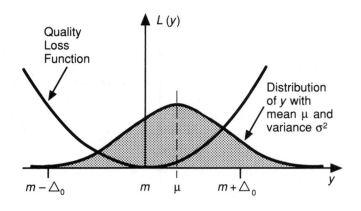

Figure 2.5   Evaluation of average quality loss.

Between these two components of quality loss, it is usually easier to eliminate the first one. Reducing the second component requires decreasing the variance, which is more difficult. Three methods of reducing variance, in order of cost effectiveness, are:

1. *Screening out bad products.* Here, products that are outside certain limits, $m \pm \Delta'$, are rejected as defective. Typically $\Delta' < \Delta_0$ so that measurement errors and product deterioration are properly taken into account. The rejected pieces are either reworked or scrapped. Because inspection, scrap, and rework are expensive, this method of reducing the variance leads to higher cost per passed product.

2. *Discovering the cause of malfunction and eliminating it.* Variance can also be reduced by discovering the cause of malfunction and eliminating it. For example, if the cause of malfunction is fluctuation in the ambient temperature, then the customer is asked to use the equipment in an air-conditioned place. If the tolerance on a particular component is identified as a major contributor to system performance, then a narrower tolerance is specified for that component. This method of reducing the variance, which is frequently used, is also expensive but usually less expensive than screening.

3. *Application of the Robust Design method.* A third method of reducing the variance is Robust Design. The method consists of making a product's performance insensitive to noise factors. As a result of applying Robust Design, in many cases the manufacturing tolerances do not have to be tightened, the product's usage environment does not have to be controlled tightly, and cheaper material or components can be used, all of which make Robust Design the most economical of the three methods. How does Robust Design do it? It exploits the inherent nonlinearity of the relationship among the product or process parameters, the noise factors, and the quality characteristics, as explained in the next section.

2.5 EXPLOITING NONLINEARITY

Usually, a product's quality characteristic is related to the various product parameters and noise factors through a complicated, nonlinear function. It is possible to find many combinations of product parameter values that can give the desired target value of the product's quality characteristic under *nominal noise conditions* (when the noise factors are exactly at their nominal values). However, due to the nonlinearity, these different product parameter combinations can give quite different variations in the quality characteristic, even when the noise factor variations are the same. The principal goal of Robust Design is to exploit the nonlinearity to find a combination of product parameter values that gives the smallest variation in the value of the quality characteristic around the desired target value.

The exploitation of nonlinearity can be understood through the following simple mathematical formulation. Let $\mathbf{x} = (x_1, x_2, \cdots, x_n)^T$ denote the noise factors and $\mathbf{z} = (z_1, z_2, \cdots, z_q)^T$ denote the product parameters (called *control factors*) whose values can be set by the designer. Suppose the following function describes the dependence of y on $\mathbf{x}$ and $\mathbf{z}$:

$$y = f(\mathbf{x}, \mathbf{z}) \ . \tag{2.11}$$

The deviation, Δy, of the quality characteristic from the target value caused by the deviations, Δx_i, of the noise factors from their respective nominal values can be approximated by the following formula:

$$\Delta y = \left[\frac{\partial f}{\partial x_1}\right]\Delta x_1 + \left[\frac{\partial f}{\partial x_2}\right]\Delta x_2 + \cdots + \left[\frac{\partial f}{\partial x_n}\right]\Delta x_n \ . \tag{2.12}$$

Further, if the deviations in the noise factors are uncorrelated, the variance, σ_y^2, of y can be expressed in terms of variances, $\sigma_{x_i}^2$, of the individual noise factors as

$$\sigma_y^2 = \left[\frac{\partial f}{\partial x_1}\right]^2 \sigma_{x_1}^2 + \left[\frac{\partial f}{\partial x_2}\right]^2 \sigma_{x_2}^2 + \cdots + \left[\frac{\partial f}{\partial x_n}\right]^2 \sigma_{x_n}^2 \ . \tag{2.13}$$

Thus, the variance σ_y^2 is a sum of the products of the variances of the noise factors, $\sigma_{x_i}^2$, and the *sensitivity coefficients*, $(\partial f/\partial x_i)^2$. The sensitivity coefficients are themselves functions of the control factor values. A *robust product* (or a *robust process*) is one for which the sensitivity coefficients are the smallest. Thus, it is obvious that for robust products we can allow wider manufacturing tolerances, lower grade components or materials, and a wider operating environment. The Robust Design method is a way of arriving at the robust products and processes efficiently.

The following example of the design of an electrical power supply circuit vividly illustrates the exploitation of nonlinearity for reducing sensitivity coefficients. The dependence of the output voltage, y, on the gain of a transistor, A, in the power supply circuit is shown in Figure 2.6. It is clear that this relationship is nonlinear. The relationship between the output voltage and the dividing resistor, B, is linear, as shown in the figure. To achieve a 110-volt target output voltage, one may choose A_1 as the transistor gain and B_1 as the dividing resistance. Suppose the transistor has ± 20 percent variation as a result of manufacturing variation, environmental variation, and drift. Then, the corresponding variation in the output voltage would be ± 5 volts as indicated in the figure. Suppose the allowed variation in the output voltage is ± 2 volts. That requirement can be accomplished by reducing the tolerance on the gain by a factor of about 2.5. This would, however, mean higher manufacturing cost.

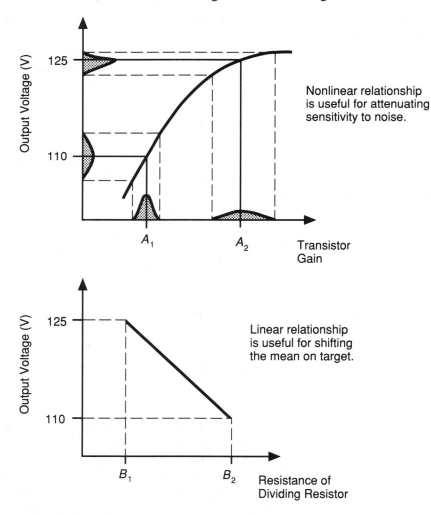

Figure 2.6   Design of a power supply circuit.

Now, consider moving the gain to A_2 where the corresponding output voltage is 125 volts. Here, for a ± 20 percent variation in the gain, the variation in the output voltage is only about ± 2 volts as shown in the figure. The mean output voltage can be brought back to the desired nominal of 110 volts by moving the resistance from B_1 to B_2. Because of linearity, this change in resistance has a negligible effect on the variation of the output voltage. Thus, we can achieve a large reduction in the variation of the output voltage by simply changing the nominal values of the transistor gain and the dividing resistor. This change, however, does not change the manufacturing cost of the circuit. Thus, by exploiting nonlinearity we can reduce the quality loss without increasing the product cost.

If the requirements on the variation of the output voltage were even tighter due to a large quality loss associated with the deviation of the output voltage from the target, the tolerance on the gain could be tightened as economically justifiable.

Thus, the variance of the output voltage can be reduced by two distinct actions:

1. Move the nominal value of gain so that the output voltage is less sensitive to the tolerance on the gain, which is noise.

2. Reduce the tolerance on the gain to control the noise.

Genichi Taguchi refers to action 1 as *parameter design* and action 2 as *tolerance design*.

Typically, no manufacturing cost increase is associated with changing the nominal values of product parameters *(parameter design)*. However, reducing tolerances *(tolerance design)* leads to higher manufacturing cost.

Managing the Economics of Quality Improvement

It is obvious from the preceding discussion that to minimize the total cost, which consists of the unit manufacturing cost and quality loss, we must first carry out parameter design. Next, during tolerance design, we should adjust the tolerances to strike an economic balance between reduction in quality loss and increase in manufacturing cost. This strategy for minimization of the total cost is a more precise statement of the fundamental principle of Robust Design described in Chapter 1.

Engineers and managers, unaware of the benefits of designing robust products and the Robust Design methodology, tend to use more costly parts, components, and manufacturing processes to improve quality without first obtaining the most benefit out of parameter design. As a result, they miss the opportunity to improve quality without increasing manufacturing cost. This leads to the misconception that higher quality always means more umc.

The average quality loss evaluation described in Section 2.4 can be used to justify the investment in quality improvement for a specific product or process. The investment consists of two parts: R&D cost associated with parameter design (this cost should be normalized by the projected sales volume) and umc associated with tolerance design. The role of the quadratic loss function and the average quality loss evaluation in managing the economics of continuous quality improvement is discussed in detail by Sullivan [S5].

Although parameter design may not increase the umc, it is not necessarily free of cost. It needs an R&D budget to explore the nonlinear effects of the various control factors. By using the techniques of orthogonal arrays and signal-to-noise ratios, which are an integral part of the Robust Design method, one can greatly improve the R&D efficiency when compared to the efficiency of the present practice of studying one control factor at a time or an ad-hoc method of finding the best values of many control factors simultaneously. Thus, by using the Robust Design method, there is potential to also reduce the total R&D cost.

2.6 CLASSIFICATION OF PARAMETERS: P DIAGRAM

A block diagram representation of a product is shown in Figure 2.7. The response of the product is denoted by y. The response could be the output of the product or some other suitable characteristic. Recall that the response we consider for the purpose of optimization in a Robust Design experiment is called a quality characteristic.

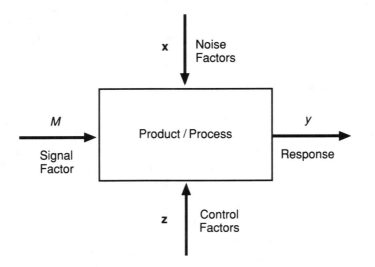

Figure 2.7   Block diagram of a product/process: P Diagram.

A number of parameters can influence the quality characteristic or response of the product. These parameters can be classified into the following three classes (note that the word *parameter* is equivalent to the word *factor* in most of Robust Design literature):

1. *Signal factors (M)*. These are the parameters set by the user or operator of the product to express the intended value for the response of the product. For example, the speed setting on a table fan is the signal factor for specifying the amount of breeze; the steering wheel angle is a signal factor that specifies the turning radius of an automobile. Other examples of signal factors are the 0 and 1 bits transmitted in a digital communication system and the original document to be copied by a photocopying machine. The signal factors are selected by the design engineer based on the engineering knowledge of the product being developed. Sometimes two or more signal factors are used in combination to express the desired response. Thus, in a radio receiver, tuning could be achieved by using the coarse and fine-tuning knobs in combination.

2. *Noise factors* (**x**). Certain parameters cannot be controlled by the designer and are called *noise factors*. Section 2.3 described three broad classes of noise factors. Parameters whose settings (also called *levels*) are difficult to control in the field or whose levels are expensive to control are also considered noise factors. The levels of the noise factors change from one unit to another, from one environment to another, and from time to time. Only the statistical characteristics (such as the mean and variance) of noise factors can be known or specified but the actual values in specific situations cannot be known. The noise factors cause the response *y* to deviate from the target specified by the signal factor *M* and lead to quality loss.

3. *Control factors* (**z**). These are parameters that can be specified freely by the designer. In fact, it is the designer's responsibility to determine the best values of these parameters. Each control factor can take multiple values, called *levels*. When the levels of certain control factors are changed, the manufacturing cost does not change; however, when the levels of others are changed, the manufacturing cost also changes. In the power supply circuit example of Section 2.5, the transistor gain and the dividing resistance are control factors that do not change the manufacturing cost. However, the tolerance of the transistor gain has a definite impact on the manufacturing cost. We will refer to the control factors that affect manufacturing cost as *tolerance factors*, whereas the other control factors simply will be called control factors.

The block diagram of Figure 2.7 can be used to represent a manufacturing process or even a business system. Identifying important responses, signal factors, noise factors, and control factors in a specific project are important tasks. In planning a Robust Design project, it is also important to recognize which control factors change the manufacturing cost and which do not. The best settings of the latter are determined through parameter design, whereas the best settings of the former are determined through tolerance design. [Sometimes, tolerance factors are also optimized during parameter design (see Chapters 10 and 11).]

Robust Design projects can be classified on the basis of the nature of the signal factor and the quality characteristic. In some problems, the signal factor takes a constant value. Such problems are called *static problems*. The other problems are called *dynamic problems*. These and other types of problems are described in Chapter 5.

Thus far in this chapter, we have described the basic principles of quality engineering, including the quadratic loss function, the exploitation of nonlinearity, and the classification of product or process parameters. All this material creates a foundation for discussing the optimization of the design of products and processes in the next section.

2.7 OPTIMIZATION OF PRODUCT AND PROCESS DESIGN

Designing a product or a manufacturing process is a complex activity. The output of the activity is a set of drawings and written specifications that specify how to make the particular product. Three essential elements of these drawings and specifications are: (a) system architecture, (b) nominal values for all parameters of the system, and (c) the tolerance or the allowable variation in each parameter. Optimizing a product or process design means determining the best architecture, the best parameter values, and the best tolerances.

Optimization Strategy for Becoming a Preferred Supplier

Consider a market where there are two suppliers for a product and the customers are capable of evaluating their quality loss. Recall that the quality loss includes the operating cost of the product as well as other losses due to the deviation of the product's function from the target. Suppose the suppliers differ in price and quality loss for their products. In such a market, the preferred supplier would be the one for whom the sum total of quality loss and price is the smallest. Depending on marketing strategy and corporate policy, a supplier can adopt one of many optimization strategies for becoming a preferred supplier. Among them, three noteworthy strategies are:

1. Minimize manufacturing cost while delivering the same quality as the competitor. Here the supplier would be able to maximize per unit profit.

2. Minimize the quality loss while keeping the manufacturing cost the same as the competitor (as judged by the price). With this strategy, the supplier can build a reputation for quality.

3. Minimize the sum of the quality loss and manufacturing cost. This is a strategy for best utilization of the sum of supplier's and customer's resources. It is most appropriate when the supplier and the customer are part of the same corporation. Also, public utility commissions are required to follow this strategy in regulating the utility companies.

Note that strategies 1 and 2 are the extreme strategies a supplier can follow to remain a preferred supplier. In between, there are infinitely many strategies, and strategy 3 is an important one among them.

Engineering Design Problem

Consider the strategy of minimizing the manufacturing cost while delivering a specified quality level. The engineering problem of optimizing a product or process design to reflect this strategy is difficult and fuzzy. First, the relationship between the numerous parameters and the response is often unknown and must be observed experimentally. Secondly, during product or process design the precise magnitudes of noise factor variations and the costs of different grades of materials, components, and tolerances are not known. For example, during product design, exact manufacturing variations are not known unless existing processes are to be used. Therefore, writing a single objective function encompassing all costs is not possible. Considering these difficulties, the following strategy has an intuitive appeal and consists of three steps: (1) concept design, (2) parameter design, and (3) tolerance design. These steps are described below.

1. *Concept design.* In this step, the designer examines a variety of architectures and technologies for achieving the desired function of the product and selects the most suitable ones for the product. Selecting an appropriate circuit diagram or a sequence of manufacturing steps are examples of concept design activity. This is a highly creative step in which the experience and skill of the designer play an important role. Usually, only one architecture or technology is selected based on the judgment of the designer. However, for highly complex products, two or three promising architectures are selected; each one is developed separately, and, in the end, the best architecture is adopted. Concept design can play an important role in reducing the sensitivity to noise factors as well as in reducing the manufacturing cost. Quality Function Deployment (QFD) and Pugh's concept selection method are two techniques that can improve the quality and productivity of the concept design step (see Clausing [C1], Sullivan [S6], Hauser and Clausing [H1], and Cohen [C4]).

2. *Parameter design.* In parameter design, we determine the best settings for the control factors that do not affect manufacturing cost, that is, the settings that minimize quality loss. Thus, we must minimize the sensitivity of the function of the product or process to all noise factors and also get the mean function on target. During parameter design, we assume wide tolerances on the noise factors and assume that low-grade components and materials would be used; that is, we fix the manufacturing cost at a low value and, under these conditions, minimize the sensitivity to noise, thus minimizing the quality loss. If at the end of parameter design the quality loss is within specifications, we have a design with the lowest cost and we need not go to the third step. However, in practice the

quality loss must be further reduced; therefore, we always have to go to the third step.

3. *Tolerance design.* In tolerance design, a trade-off is made between reduction in the quality loss due to performance variation and increase in manufacturing cost; that is, we selectively reduce tolerances and selectively specify higher-grade material (note that these are all tolerance factors) in the order of their cost effectiveness. Tolerance design should be performed only after sensitivity to noise has been minimized through parameter design. Otherwise, to achieve the desired low value of quality loss, we would have to specify unnecessarily higher-grade materials and components leading to higher manufacturing cost. Sometimes the variation in a product's response can be reduced by adding a suitable compensation mechanism, such as feedback control. Of course, this leads to higher product cost. Thus, the inclusion of a compensation mechanism should be considered as a tolerance factor to be optimized along with the component tolerances.

Many Japanese companies do an excellent job of minimizing sensitivity to noise by using the Robust Design method. As a result, they can manufacture higher quality products at a lower cost. Many American companies, unaware of the Robust Design method, depend heavily on tolerance design and concept design to improve quality. Relying on tolerance design makes products more expensive to manufacture, and relying on improved concept design requires achieving breakthroughs which are difficult to schedule and, hence, lead to longer development time.

Robust Design and its associated methodology focus on parameter design. A full treatment of concept design is beyond the scope of this book. Tolerance design is discussed briefly in Chapters 8 and 11 with case studies.

At the beginning of this section, we described three optimization strategies for becoming a preferred supplier. Each of these strategies can be realized through the concept design, parameter design, and tolerance design. The only difference is in the tolerance design step in which the "stopping rule" is different for the selective process of specifying higher-grade material and components. To maximize per unit profit, we should stop the selective specification of higher-grade material as soon as the quality loss for the product equals that of our competitors. To gain a reputation for quality, we should stop as soon as our manufacturing cost equals the competitor's manufacturing cost. To maximize the use of societal resources, we should stop as soon as the marginal increase in manufacturing cost equals the marginal reduction in quality loss. In all these cases, parameter design is an essential step for gaining the most benefit of a given concept design.

The engineering design activity for complicated products is often organized in the following hierarchy: (1) design of the overall system, (2) subsystem design, and (3) component design. The steps of concept design, parameter design, and tolerance design can be applied in each of the three hierarchical levels.

2.8 ROLE OF VARIOUS QUALITY CONTROL ACTIVITIES

The goal of this section is to delineate the major quality control activities during the various stages of the life cycle of a product and to put in perspective the role of the Robust Design methodology. Once a decision to make a product has been made, the life cycle of that product has four major stages:

1. Product design

2. Manufacturing process design

3. Manufacturing

4. Customer usage

The quality control activities in each of these stages are listed in Table 2.1. The quality control activities in product and process design are called *off-line quality control*, whereas the quality control activities in manufacturing are called *on-line quality control*. For customer usage, quality control activities involve warranty and service.

Product Design

During product design, one can address all three types of noise factors (external, unit-to-unit variation, and deterioration), thus making it the most important stage for improving quality and reducing the unit manufacturing cost. Parameter design during this stage reduces sensitivity to all three types of noise factors and, thus, gives us the following benefits:

- The product can be used in a wide range of environmental conditions, so the product's operating cost is lower.

- The use of costly compensation devices, such as temperature-compensation circuits and airtight encapsulation, can be avoided, thus making the product simpler and cheaper.

- Lower-grade components and materials can be used.

- The benefits that can be derived from making manufacturing process design robust are also realized through parameter design during product design.

Manufacturing Process Design

During manufacturing process design, we cannot reduce the effects of either the external noise factors or the deterioration of the components on the product's performance in the field. This can be done only through product design and selection of material and components. However, the unit-to-unit variation can be reduced through

TABLE 2.1 QUALITY CONTROL ACTIVITIES DURING VARIOUS PRODUCT REALIZATION STEPS

Product Realization Step	Quality Control Activity	Ability to Reduce Effect of Noise Factors			Comments
		External	Unit to Unit	Deterior-ation	
Product design	a) Concept design	Yes	Yes	Yes	Involves innovation to reduce sensitivity to all noise factors.
	b) Parameter design	Yes	Yes	Yes	Most important step for reducing sensitivity to all noise factors. Uses Robust Design method.
	c) Tolerance design	Yes	Yes	Yes	Method for selecting most economical grades of materials, components and manufacturing equipment, and operating environment for the product.
Manufacturing process design	a) Concept design	No	Yes	No	Involves innovation to reduce unit-to-unit variation.
	b) Parameter design	No	Yes	No	Important for reducing sensitivity of unit-to-unit variation to manufacturing variations.
	c) Tolerance design	No	Yes	No	Method for determining tolerances on manufacturing process parameters.
Manufacturing	a) Detection and correction	No	Yes	No	Method of detecting problems when they occur and correcting them.
	b) Feedforward control	No	Yes	No	Method of compensating for known problems.
	c) Screening	No	Yes	No	Last alternative, useful when process capability is poor.
Customer usage	Warranty and Repair	No	No	No	

Source: Adapted from G. Taguchi, "Off-line and On-line Quality Control System," International Conference on Quality Control, Tokyo, Japan, 1978.

process design because parameter design during process design reduces the sensitivity of the unit-to-unit variation to the various noise factors that affect the manufacturing process.

For some products the deterioration rate may depend on the design of the manufacturing process. For example, in microelectronics, the amount of impurity has a direct relationship with the deterioration rate of the integrated circuit and it can be controlled through process design. For some mechanical parts, the surface finish can determine the wear-out rate and it too can be controlled by the manufacturing process design. Therefore, it is often said that the manufacturing process design can play an important role in controlling the product deterioration. However, it is not the same thing as making the product's performance insensitive to problems, such as impurities or surface finish. Reducing the sensitivity of the product's performance can only be done during product design. In the terminology of Robust Design, we consider the problems of impurities or surface finish as a part of manufacturing variation (unit-to-unit variation) around the target.

The benefits of parameter design in process design are:

- The expense and time spent in final inspection and on rejects can be reduced greatly.

- Raw material can be purchased from many sources and the expense of incoming material inspection can be reduced.

- Less expensive manufacturing equipment can be used.

- Wider variation in process conditions can be permitted, thus reducing the need and expense of on-line quality control (process control).

Manufacturing

No matter how well a manufacturing process is designed, it will not be perfect. Therefore, it is necessary to have on-line quality control during daily manufacturing so that the unit-to-unit variation can be minimized as justified by manufacturing economics. There are three major types of on-line quality control activities (see Taguchi [T3] for detailed discussion of on-line quality control):

1. *Detection and correction.* Here the goal is to recognize promptly the breakdown of a machine or a piece of equipment, a change in raw material characteristic, or an operation error which has a consistent effect on the process. This is accomplished by periodically observing either the process conditions or product characteristics. Once such a deviation is recognized, appropriate corrective action is taken to prevent the future units from being off-target. The detection and correction method is a way of balancing the customer's quality loss resulting from unit-to-unit variation against the manufacturer's operating expenses, including the cost of periodic testing and correction of problems. Thus, it includes activities such as preventive maintenance and buying better test equipment. Statistical process control (SPC) techniques are often used for detecting process problems (see Grant [G2], Duncan [D5], and Feigenbaum [F1]).

2. *Feedforward control.* Here, the goal is to send information about errors or problems discovered in one step to the next step in the process so that variation can be reduced. Consider a situation where, by mistake, a 200 ASA film is exposed with a 100 ASA setting on the camera. If that information is passed on to a film

developing organization, the effect of the error in exposure can be reduced by adjusting the developing parameters. Similarly, by measuring the properties of incoming material or informing the subsequent manufacturing step about the problems discovered in an earlier step, unit-to-unit variation in the final product can be reduced. This is the essence of feedforward control.

3. *Screening.* Here, the goal is to stop defective units from being shipped. In certain situations, the manufacturing process simply does not have adequate capability—that is, even under nominal operating conditions the process produces a large number of defective products. Then, *as the last alternative,* all units produced can still be measured and the defective ones discarded or repaired to prevent shipping them to customers. In electronic component manufacturing, it is common to *burn-in* the components (subject the components to normal or high stress for a period of time) as a method for screening out the bad components.

Customer Usage

With all the quality control efforts in product design, process design, and manufacturing, some defective products may still get shipped to the customer. The only way to prevent further damage to the manufacturer's reputation for quality is to provide field service and compensate the customer for the loss caused by the defective product.

2.9 SUMMARY

- Quality engineering is concerned with reducing both the quality loss, which is the cost incurred after the sale of a product, and the unit manufacturing cost (umc).

- Fraction defective is often an incomplete and misleading measure of quality. It connotes that all products that meet the specification limits are equally good, while those outside the specification limits are bad. However, in practice the quality becomes progressively worse as the product's response deviates from the target value.

- The quadratic loss function is a simple and meaningful function for approximating the quality loss in most situations. The three most common variations of the quadratic loss function are:

    1. *Nominal-the-best type*: $L(y) = \dfrac{A_0}{\Delta_0^2} (y - m)^2$

    2. *Smaller-the-better type*: $L(y) = \dfrac{A_0}{\Delta_0^2} y^2$

3. *Larger-the-better type*: $L(y) = A_0 \Delta_0^2 \left[\dfrac{1}{y^2} \right]$

In the formulae above, Δ_0 is the functional limit and A_0 is the loss incurred at the functional limit. The target values of the response (or the quality characteristic) for the three cases are m, 0, and ∞, respectively.

- A product response that is observed for the purpose of evaluating the quality loss or optimizing the product design is called a *quality characteristic*. The parameters (also called *factors*) that influence the quality characteristic can be classified into three classes:

  1. Signal factors are the factors that specify the intended value of the product's response.

  2. Noise factors are the factors that cannot be controlled by the designer. Factors whose settings are difficult or expensive to control are also called noise factors. The noise factors themselves can be divided into three broad classes: (1) external (environmental and load factors), (2) unit-to-unit variation (manufacturing nonuniformity), and (3) deterioration (wear-out, process drift).

  3. Control factors are the factors that can be specified freely by the designer. Their settings (or levels) are selected to minimize the sensitivity of the product's response to all noise factors. Control factors that also affect the product's cost are called tolerance factors.

- A robust product or a robust process is one whose response is least sensitive to all noise factors. A product's response depends on the values of the control and noise factors through nonlinear function. We exploit the nonlinearity to achieve robustness.

- The three major steps in designing a product or a process are:

  1. Concept design—selection of product architecture or process technology.

  2. Parameter design—selection of the optimum levels of the control factors to maximize robustness.

  3. Tolerance design—selection of the optimum values of the tolerance factors (material type, tolerance limits) to balance the improvement in quality loss against the increase in the umc.

- Quality improvement through concept design needs breakthroughs, which are difficult to schedule. Parameter design improves quality without increasing the umc. It can be performed systematically by using orthogonal arrays and the signal-to-noise ratios, which is the most inexpensive way to improve quality.

Tolerance design should be undertaken only after parameter design is done; otherwise, the product's umc may turn out to be unnecessarily high. Japanese companies have gained huge quality and cost advantage by emphasizing parameter design. The Robust Design methodology focuses on how to perform parameter design efficiently.

- Depending on marketing strategy and corporate policy, a supplier can adopt one of many optimization strategies for becoming a preferred supplier. Among them, three noteworthy strategies are: (1) minimize manufacturing cost while delivering the same quality as the competition, (2) minimize the quality loss while keeping the manufacturing cost the same as the competition, and (3) minimize the sum of the quality loss and the manufacturing cost. Regardless of which optimization strategy is adopted, one must first perform parameter design.

- The life cycle of a product has four major stages: (1) product design, (2) manufacturing process design, (3) manufacturing, and (4) customer usage. Quality control activities during product and process design are called *off-line* quality control, while those in manufacturing are called *on-line* quality control. Warranty and service are the ways for dealing with quality problems during customer usage.

- A product's sensitivity to all three types of noise factors can be reduced during product design, thus making product design the most important stage for improving quality and reducing umc. The next important step is manufacturing process design through which the unit-to-unit variation (and some aspects of deterioration) can be reduced along with the umc. During manufacturing, the unit-to-unit variation can be further reduced, but with less cost effectiveness than during manufacturing process design.

Chapter 3

MATRIX EXPERIMENTS USING ORTHOGONAL ARRAYS

A *matrix experiment* consists of a set of experiments where we change the settings of the various product or process parameters we want to study from one experiment to another. After conducting a matrix experiment, the data from all experiments in the set taken together are analyzed to determine the effects of the various parameters. Conducting matrix experiments using special matrices, called *orthogonal arrays*, allows the effects of several parameters to be determined efficiently and is an important technique in Robust Design.

This chapter introduces the technique of matrix experiments based on orthogonal arrays through a simulated example of a chemical vapor deposition (CVD) process. In particular, it focuses on the analysis of data collected from matrix experiments and the benefits of using orthogonal arrays. The engineering issues involved in planning and conducting matrix experiments are discussed in Chapter 4, and the techniques of constructing orthogonal arrays are discussed in Chapter 7.

This chapter consists of six sections:

- Section 3.1 describes the matrix experiment and the concept of orthogonality.

- Section 3.2 shows how to analyze the data from matrix experiments to determine the effects of the various parameters or factors. One of the benefits of using orthogonal arrays is the simplicity of data analysis. The effects of the various factors can be determined by computing simple averages, an approach that has an intuitive appeal. The estimates of the factor effects are then used to determine the optimum factor settings.

- Section 3.3 presents a model, called *additive model*, for the factor effects and demonstrates the validity of using simple averages for estimating the factor effects.

- Section 3.4 describes *analysis of variance* (ANOVA), a useful technique for estimating error variance and for determining the relative importance of various factors. Here we also point out the analogy between ANOVA and *Fourier series decomposition*.

- Section 3.5 discusses the use of the additive model for prediction and diagnosis, as well as the concept of interaction and the detection of the presence of interaction through diagnosis.

- Section 3.6 summarizes the important points of this chapter.

In statistical literature, matrix experiments are called *designed experiments* and the individual experiments in a matrix experiment are sometimes called *runs* or *treatments*. Settings are also referred to as *levels* and parameters as *factors*. The theory behind the analysis of matrix experiments can be found in many text books on experimental design, including Hicks [H2]; Box, Hunter, and Hunter [B3]; Cochran and Cox [C3]; John [J2]; and Daniel [D1].

3.1 MATRIX EXPERIMENT FOR A CVD PROCESS

Consider a project where we are interested in determining the effect of four process parameters: temperature (A), pressure (B), settling time (C), and cleaning method (D) on the formation of certain surface defects in a chemical vapor deposition (CVD) process. Suppose for each parameter three settings are chosen to cover the range of interest.

The factors and their chosen levels are listed in Table 3.1. The *starting levels* (levels before conducting the matrix experiment) for the four factors, identified by an underscore in the table, are: T_0°C temperature, P_0 mtorr pressure, t_0 minutes of settling time, and no cleaning. The alternate levels for the factors are as shown in the table; for example, the two alternate levels of temperature included in the study are (T_0-25)°C and (T_0+25)°C. These factor levels define the *experimental region* or the region of interest. Our goal for this project is to determine the best setting for each parameter so that the surface defect formation is minimized.

The matrix experiment selected for this project is given in Table 3.2. It consists of nine individual experiments corresponding to the nine rows. The four columns of the matrix represent the four factors as indicated in the table. The entries in the matrix represent the levels of the factors. Thus, experiment 1 is to be conducted with each factor at the first level. Referring to Table 3.1, we see that the factor levels for experiment 1 are (T_0-25)°C temperature, (P_0-200) mtorr pressure, t_0 minute settling time, and no cleaning.   Similarly, by referring to Tables 3.2 and 3.1, we see that experiment 4 is to be conducted at level 2 of temperature (T°C), level 1 of pressure ($P_0 - 200$ mtorr), level 2 of settling time ($t_0 + 8$ minutes), and level 3 of cleaning method (CM_3). The settings of experiment 4 can also be referred to concisely as $A_2\ B_1\ C_2\ D_3$.

TABLE 3.1 FACTORS AND THEIR LEVELS

		Levels*	
Factor	**1**	**2**	**3**
A. Temperature (°C)	$T_0 - 25$	$\underline{T_0}$	$T_0 + 25$
B. Pressure (mtorr)	$P_0 - 200$	$\underline{P_0}$	$P_0 + 200$
C. Settling time (min)	$\underline{t_0}$	$t_0 + 8$	$t_0 + 16$
D. Cleaning method	$\underline{\text{None}}$	CM_2	CM_3

* The starting level for each factor is identified by an underscore.

TABLE 3.2 MATRIX EXPERIMENT

	Column Number and Factor Assigned				Observation*
Expt. No.	**1 Temperature (A)**	**2 Pressure (B)**	**3 Settling Time (C)**	**4 Cleaning Method (D)**	**η (dB)**
1	1	1	1	1	$\eta_1 = -20$
2	1	2	2	2	$\eta_2 = -10$
3	1	3	3	3	$\eta_3 = -30$
4	2	1	2	3	$\eta_4 = -25$
5	2	2	3	1	$\eta_5 = -45$
6	2	3	1	2	$\eta_6 = -65$
7	3	1	3	2	$\eta_7 = -45$
8	3	2	1	3	$\eta_8 = -65$
9	3	3	2	1	$\eta_9 = -70$

* $\eta = -10 \log_{10}$ (mean square surface defect count).

The matrix experiment of Table 3.2 is the standard orthogonal array L_9 of Taguchi and Wu [T7]. As the name suggests, the columns of this array are mutually orthogonal. Here, orthogonality is interpreted in the combinatoric sense—that is, for any pair of columns, all combinations of factor levels occur and they occur an equal number of times. This is called the *balancing property* and it implies orthogonality. A more formal mathematical definition of orthogonality of a matrix experiment is given in Appendix A at the end of this book. In this matrix, for each pair of columns, there exist $3 \times 3 = 9$ possible combinations of factor levels, and each combination occurs precisely once. For columns 1 and 2, the nine possible combinations of factor levels, namely the combinations (1,1), (1,2), (1,3), (2,1), (2,2), (2,3), (3,1), (3,2), and (3,3), occur in experiments (or rows) 1, 2, 3, 4, 5, 6, 7, 8, and 9, respectively. In all, six pairs of columns can be formed from the four columns. We ask the reader to verify orthogonality for at least a few of these pairs.

A common method of finding the frequency response function of a dynamic system to time varying input is to observe the output of the system for sinusoidal inputs of different frequencies, one frequency at a time. Another approach is to use an input consisting of several sinusoidal frequencies and observe the corresponding output. Fourier analysis is then used to determine the gain and phase for each frequency. Conducting matrix experiments with several factors is analogous to the use of multifrequency input for finding the frequency response function. The analogy of matrix experiments with Fourier analysis is discussed further in Section 3.5.

There exists a large variety of industrial experiments. Each experiment has a different number of factors. Some factors have two levels, some three levels, and some even more. The following sections discuss the analysis of such experiments and demonstrate the benefits of using orthogonal arrays to plan matrix experiments. A number of standard orthogonal arrays and the techniques of constructing orthogonal arrays to suit specific projects are described in Chapter 7.

3.2 ESTIMATION OF FACTOR EFFECTS

Suppose for each experiment we observe the surface defect count per unit area at three locations each on three *silicon wafers* (thin disks of silicon used for making VLSI circuits) so that there are nine observations per experiment. We define by the following formula a summary statistic, η_i, for experiment i:

$$\eta_i = -10 \log_{10} \text{ (mean square defect count for experiment } i)$$

where the mean square refers to the average of the squares of the nine observations in experiment i. We refer to the η_i calculated using the above formula as the observed η_i. Let the observed η_i for the nine experiments be as shown in Table 3.2. Note that the objective of minimizing surface defects is equivalent to maximizing η. The summary statistic η is called signal-to-noise (S/N) ratio. The rationale for using η as the objective function is discussed in Chapter 5.

Let us now see how to estimate the effects of the four process parameters from the observed values of η for the nine experiments. First, the overall mean value of η for the experimental region defined by the factor levels in Table 3.1 is given by

$$m = \frac{1}{9} \sum_{i=1}^{9} \eta_i$$

$$= \frac{1}{9} \left[\eta_1 + \eta_2 + \cdots + \eta_9 \right] . \tag{3.1}$$

By examining columns 1, 2, 3, and 4 of the orthogonal array in Table 3.2, observe that all three levels of every factor are equally represented in the nine experiments. Thus, m is a balanced overall mean over the entire experimental region.

The *effect of a factor level* is defined as the deviation it causes from the overall mean. Let us examine how the experimental data can be used to evaluate the effect of temperature at level A_3. Temperature was at level A_3 for experiments 7, 8, and 9. The average S/N ratio for these experiments, which is denoted by m_{A_3}, is given by

$$m_{A_3} = \frac{1}{3} \left[\eta_7 + \eta_8 + \eta_9 \right] . \tag{3.2}$$

Thus, the effect of temperature at level A_3 is given by $(m_{A_3} - m)$. From Table 3.2, observe that for experiments 7, 8, and 9, the pressure level takes values 1, 2, and 3, respectively. Similarly, for these three experiments, the levels of settling time and cleaning method also take values 1, 2, and 3. So the quantity m_{A_3} represents an average η when the temperature is at level A_3 where the averaging is done in a balanced manner over all levels of each of the other three factors.

The average S/N ratio for levels A_1 and A_2 of temperature, as well as those for the various levels of the other factors, can be obtained in a similar way. Thus, for example,

$$m_{A_2} = \frac{1}{3} \left[\eta_4 + \eta_5 + \eta_6 \right] \tag{3.3}$$

represents the average S/N ratio for temperature at level 2, and

$$m_{B_2} = \frac{1}{3} \left[\eta_2 + \eta_5 + \eta_8 \right] \tag{3.4}$$

is the average S/N ratio for pressure at level B_2. Because the matrix experiment is based on an orthogonal array, all the level averages possess the same balancing property described for m_{A_3}.

By taking the numerical values of η listed in Table 3.2, the average η for each level of the four factors can be obtained as listed in Table 3.3. These averages are shown graphically in Figure 3.1. They are separate effects of each factor and are commonly called *main effects*. The process of estimating the factor effects discussed above is sometimes called *analysis of means* (ANOM).

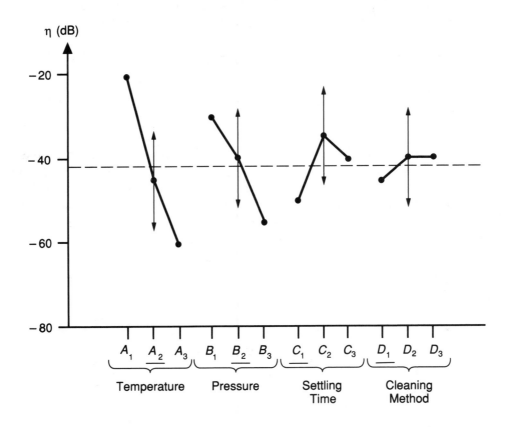

Figure 3.1   Plots of factor effects.   Underscore indicates starting level. Two-standard-deviation confidence limits are also shown for the middle level.

TABLE 3.3 AVERAGE η BY FACTOR LEVELS (dB)

Factor	Level 1	Level 2	Level 3
A. Temperature	−20*	−45	−60
B. Pressure	−30*	−40	−55
C. Settling time	−50	−35*	−40
D. Cleaning method	−45	−40*	−40*

* Overall mean = −41.67 dB. Starting level is identified by an underscore, and the optimum level is identified by *.

Selecting Optimum Factor Levels

A primary goal in conducting a matrix experiment is to *optimize* the product or process design—that is, to determine the best or the *optimum level* for each factor. The optimum level for a factor is the level that gives the highest value of η in the experimental region. The estimated main effects can be used for this purpose provided the variation of η as a function of the factor levels follows the *additive model* described in the next section. How to ensure that the additive model holds in a specific project is a crucial question, which is addressed later in this chapter and also at various places in the rest of the book as the appropriate situations arise.

Recall that our goal in the CVD project is to minimize the surface defect count. Since −log is a monotone decreasing function, it implies that we should maximize η. Note that η = −20 is preferable to η = −45, because −20 is greater than −45. From Figure 3.1 we can determine the optimum level for each factor as the level that has the highest value of η. Thus, the best temperature setting is A_1, the best pressure is B_1, the best settling time is C_2, and the best cleaning method could be D_2 or D_3, since the average η for both D_2 and D_3 is −40 dB. Based on the matrix experiment, we can conclude that the settings $A_1 B_1 C_2 D_2$ and $A_1 B_1 C_2 D_3$ would give the highest η or the lowest surface defect count.

The predicted best settings need not correspond to one of the rows in the matrix experiment. In fact, often they do not correspond as is the case in the present example. Also, typically, the value of η realized for the predicted best settings is better than the best among the rows of the matrix experiment.

3.3 ADDITIVE MODEL FOR FACTOR EFFECTS

In the preceding section, we used simple averaging to estimate factor effects. The same nine observations (η_1, η_2, $\cdots$, η_9) are grouped differently to estimate the factor effects. Also, the optimum combination of settings was determined by examining the effect of each factor separately. Justification for this simple procedure comes from

- Use of the additive model as an approximation

- Use of an orthogonal array to plan the matrix experiment

We now examine the additive model. The relationship between η and the process parameters A, B, C, and D can be quite complicated. Empirical determination of this relationship can, therefore, turn out to be quite expensive. However, in most situations, when η is chosen judiciously, the relationship can be approximated adequately by the following additive model:

$$\eta\,(A_i,\; B_j,\; C_k,\; D_l) = \mu + a_i + b_j + c_k + d_l + e. \tag{3.5}$$

In the above equation, μ is the overall mean—that is, the mean value of η for the experimental region; the deviation from μ caused by setting factor A at level A_i is a_i; the terms b_j, c_k and d_l represent similar deviations from μ caused by the settings B_j, C_k and D_l of factors B, C, and D, respectively; and e stands for the error. Note that *by error we imply the error of the additive approximation plus the error in the repeatability of measuring η for a given experiment.*

An additive model is also referred to as a *superposition model* or a *variables separable model* in engineering literature. Note that superposition model implies that the total effect of several factors (also called variables) is equal to the sum of the individual factor effects. It is possible for the individual factor effects to be linear, quadratic, or of higher order. However, in an additive model cross product terms involving two or more factors are not allowed.

By definition a_1, a_2, and a_3 are the deviations from μ caused by the three levels of factor A. Thus,

$$a_1 + a_2 + a_3 = 0\;. \tag{3.6}$$

Similarly,

$$\left. \begin{aligned} b_1 + b_2 + b_3 &= 0 \\ c_1 + c_2 + c_3 &= 0 \\ d_1 + d_2 + d_3 &= 0 \end{aligned} \right\} . \tag{3.7}$$

It can be shown that the averaging procedure of Section 3.2 for estimating the factor effects is equivalent to fitting the additive model, defined by Equations (3.5), (3.6), and (3.7), by the *least squares method*. This is a consequence of using an orthogonal array to plan the matrix experiment.

Now, consider Equation (3.2) for the estimation of the effect of setting temperature at level 3:

$$m_{A_3} = \frac{1}{3} (\eta_7 + \eta_8 + \eta_9)$$

$$= \frac{1}{3} \Big[(\mu + a_3 + b_1 + c_3 + d_2 + e_7)$$

$$+ (\mu + a_3 + b_2 + c_1 + d_3 + e_8)$$

$$+ (\mu + a_3 + b_3 + c_2 + d_1 + e_9) \Big]$$

$$= \frac{1}{3} (3\mu + 3a_3) + \frac{1}{3} (b_1 + b_2 + b_3) + \frac{1}{3} (c_1 + c_2 + c_3)$$

$$+ \frac{1}{3} (d_1 + d_2 + d_3) + \frac{1}{3} (e_7 + e_8 + e_9)$$

$$= (\mu + a_3) + \frac{1}{3} (e_7 + e_8 + e_9) . \tag{3.8}$$

Note that the terms corresponding to the effects of factors B, C and D drop out because of Equation (3.7). Thus, m_{A_3} is an estimate of $(\mu + a_3)$.

Furthermore, the error term in Equation (3.8) is an average of three error terms. Suppose σ_e^2 is the average variance for the error terms $e_1, e_2, \cdots, e_9$. Then the error variance for the estimate m_{A_3} is approximately $(1/3)\sigma_e^2$. (Note that in computing the error variances of the estimate m_{A_3} and other estimates in this chapter, we treat the individual error terms as independent random variables with zero mean and variance σ_e^2. In reality, this is only an approximation because the error terms include the error of the additive approximation so that the error terms are not strictly independent random variables with zero mean. This approximation is adequate because the error variance is used for only qualitative purposes.) This represents a 3-fold reduction in error variance compared to conducting a single experiment at the setting A_3 of factor A.

Substituting Equation (3.5) in Equation (3.3) verifies that m_{A_2} estimates $\mu + a_2$ with error variance $(1/3)\sigma_e^2$. Similarly, substituting Equation (3.5) in Equation (3.4) shows that m_{B_2} estimates $\mu + b_2$ with error variance $(1/3)\sigma_e^2$. It can be verified that similar relationships hold for the estimation of the remaining factor effects.

The term *replication number* is used to refer to the number of times a particular factor level is repeated in an orthogonal array. The error variance of the average effect for a particular factor level is smaller than the error variance of a single experiment by a factor equal to its replication number. To obtain the same accuracy of the factor level averages, we would need a much larger number of experiments if we were to use the traditional approach of studying one factor at a time. For example, we would have to conduct $3 \times 3 = 9$ experiments to estimate the average η for three levels of temperature alone (three repetitions each for the three levels), while keeping other factors fixed at certain levels, say, B_1, C_1, D_1.

We may then fix temperature at its best setting and experiment with levels B_2 and B_3 of pressure. This would need $3 \times 2 = 6$ additional experiments. Continuing in this manner, we can study the effects of factors C and D by performing $2 \times 6 = 12$ additional experiments. Thus, we would need a total of $9 + 3 \times 6 = 27$ experiments to study the four factors, one at a time. Compare this to only nine experiments needed for the orthogonal array based matrix experiment to obtain the same accuracy of the factor level averages.

Another common approach to finding the optimum combination of factor levels is to conduct a *full factorial experiment*—that is, conduct experiments under all combinations of factor levels. In the present example, it would mean conducting experiments under $3^4 = 81$ distinct combinations of factor levels, which is much larger than the nine experiments needed for the matrix experiment. When the additive model [Equation (3.5)] holds, it is obviously unnecessary to experiment with all combinations of factor levels. Fortunately, in most practical situations the additive model provides an excellent approximation. The additivity issue is discussed in much detail in Chapters 5 and 6.

Conducting matrix experiments using orthogonal arrays has another statistical advantage. If the errors, e_i, are independent with zero mean and equal variance, then the estimated factor effects are mutually uncorrelated. Consequently, the best level of each factor can be determined separately.

In order to preserve the benefits of using an orthogonal array, it is important that all experiments in the matrix be performed. If experiments corresponding to one or more rows are not conducted, or if their data are missing or erroneous, the balancing property and, hence, the orthogonality is lost. In some situations, incomplete matrix experiments can give useful results, but the analysis of such experiments is complicated. (Statistical techniques used for analyzing such data are regression analysis and linear models; see Draper and Smith [D4].) Thus, we recommend that any missing experiments be performed to complete the matrix.

3.4 ANALYSIS OF VARIANCE

Different factors affect the surface defect formation to a different degree. The relative magnitude of the factor effects could be judged from Table 3.3, which gives the average η for each factor level. A better feel for the relative effect of the different factors can be obtained by the decomposition of variance, which is commonly called analysis of variance (ANOVA). ANOVA is also needed for estimating the error variance for the factor effects and variance of the prediction error.

Analogy with Fourier Analysis

An important reason for performing Fourier analysis of an electrical signal is to determine the power in each harmonic to judge the relative importance of the various harmonics. The larger the amplitude of a harmonic, the larger the power is in it and the more important it is in describing the signal. Similarly, an important purpose of ANOVA is to determine the relative importance of the various factors. In fact, there is a strong analogy between ANOVA and the decomposition of the power of an electrical signal into different harmonics:

- The nine observed values of η are analogous to the observed signal.

- The sum of squared values of η is analogous to the power of the signal.

- The overall mean η is analogous to the *dc* part of the signal.

- The four factors are like four harmonics.

- The columns in the matrix experiment are orthogonal, which is analogous to the orthogonality of the different harmonics.

The analogy between the Fourier analysis of the power of an electrical signal and ANOVA is displayed in Figure 3.2. The experiments are arranged along the horizontal axis like time. The overall mean is plotted as a straight line like a dc component. The effect of each factor is displayed as an harmonic. The level of factor A for experiments 1, 2, and 3 is A_1. So, the height of the wave for A is plotted as m_{A_1} for these experiments. Similarly, the height of the wave for experiments 4, 5, and 6 is m_{A_2}, and the height for experiments 7, 8, and 9 is m_{A_3}. The waves for the other factors are also plotted similarly. By virtue of the additive model [Equation (3.5)], the observed η for any experiment is equal to the sum of the height of the overall mean and the deviation from mean caused by the levels of the four factors. By referring to the waves of the different factors shown in Figure 3.2 it is clear that factors A, B, C, and D are in the decreasing order of importance. Further aspects of the analogy are discussed in the rest of this section.

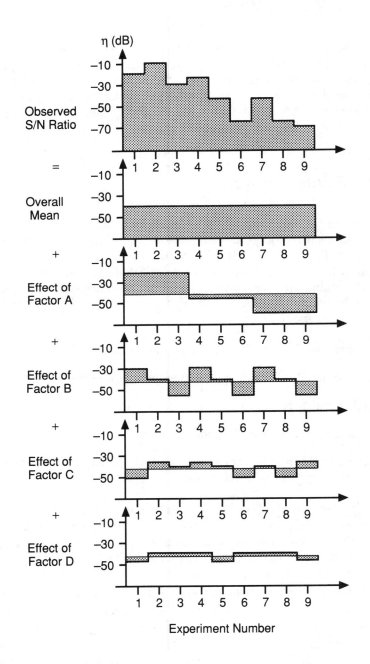

Figure 3.2   Orthogonal decomposition of the observed S/N ratio.

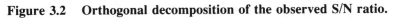

Computation of Sum of Squares

The sum of the squared values of η is called *grand total sum of squares*. Thus, we have

$$\text{Grand total sum of squares} = \sum_{i=1}^{9} \eta_i^2$$

$$= (-20)^2 + (-10)^2 + \cdots + (-70)^2$$

$$= 19{,}425 \ (\text{dB})^2 \ .$$

The grand total sum of squares is analogous to the total signal power in Fourier analysis. It can be decomposed into two parts—*sum of squares due to mean* and *total sum of squares* which are defined as follows:

$$\text{Sum of squares due to mean} = (\text{number of experiments}) \times m^2$$

$$= 9 \ (41.67)^2$$

$$= 15{,}625 \ (\text{dB})^2 \ .$$

$$\text{Total sum of squares} = \sum_{i=1}^{9} (\eta_i - m)^2$$

$$= (-20 - 41.67)^2 + (-10 - 41.67)^2 + \cdots$$

$$+ (-70 - 41.67)^2$$

$$= 3{,}800 \ (\text{dB})^2 \ .$$

The sum of squares due to mean is analogous to the *dc* power of the signal and the total sum of squares is analogous to the *ac* power of the signal in Fourier analysis. Because m is the average of the nine η_i values, we have the following algebraic identity:

$$\sum_{i=1}^{9} (\eta_i - m)^2 = \sum_{i=1}^{9} \eta_i^2 - 9m^2$$

which can also be written as

$$\text{Total sum of squares} = (\text{grand total sum of squares})$$

$$- (\text{sum of squares due to mean}) \, .$$

The above equation is analogous to the fact from Fourier analysis that the *ac* power is equal to the difference between the total power and the *dc* power of the signal.

The sum of squares due to factor A is equal to the total squared deviation of the wave for factor A from the line representing the overall mean. There are three experiments each at levels A_1, A_2, and A_3. Consequently,

Sum of squares due to factor A

$$= 3(m_{A_1} - m)^2 + 3(m_{A_2} - m)^2 + 3(m_{A_3} - m)^2$$

$$= 3(-20 + 41.67)^2 + 3(-45 + 41.67)^2 + 3(-60 + 41.67)^2$$

$$= 2450 \; (\text{dB})^2 \, .$$

Proceeding along the same lines, we can show that the sum of squares due to factors B, C, and D are, respectively, 950, 350, and 50 $(\text{dB})^2$. These sums of squares values are tabulated in Table 3.4. The sums of squares values due to various factors are analogous to the power in various harmonics, and are a measure of the relative importance of the factors in changing the values of η.

Thus, factor A explains a major portion of the total variation of η. In fact, it is responsible for $(2450/3800) \times 100 = 64.5$ percent of the variation of η. Factor *B* is responsible for the next largest portion, namely 25 percent; and factors C and D together are responsible for only a small portion, a total of 10.5 percent, of the variation in η.

Knowing the factor effects (that is, knowing the values of a_i, b_j, c_k, and d_l), we can use the additive model given by Equation (3.5) to calculate the error term e_i for each experiment i. The sum of squares due to error is the sum of the squares of the error terms. Thus we have,

$$\text{Sum of squares due to error} = \sum_{i=1}^{9} e_i^2 \, .$$

In the present case study, the total number of model parameters (μ, a_1, a_2, a_3, b_1, b_2, etc.) is 13; the number of constraints, defined by Equations (3.6) and (3.7) is 4. The number of model parameters minus the number of constraints is equal to the number of experiments. Hence, the error term is identically zero for each experiment. Hence, the sum of squares due to error is also zero. Note that this need not be the situation with all matrix experiments.

TABLE 3.4 ANOVA TABLE FOR η

Factor/Source	Degrees of Freedom	Sum of Squares	Mean Square	F
A. Temperature	2	2450	1225	12.25
B. Pressure	2	950	475	4.75
C. Settling time	2	350*	175	
D. Cleaning method	2	50*	25	
Error	0	0	–	
Total	8	3800		
(Error)	(4)	(400)	(100)	

* Indicates sum of squares added together to estimate the pooled
   error sum of squares indicated by parentheses. F ratio is calculated
   by using the pooled error mean square.

Relationship Among the Various Sums of Squares

The orthogonality of the matrix experiment implies the following relationship among the various sums of squares:

(Total sum of squares) = (sum of the sums of squares

due to various factors)

+ (sum of squares due to error) .   (3.9)

Equation (3.9) is analogous to Parseval's equation for the decomposition of the power of a signal into power in different harmonics. Equation (3.9) is often used for calculating the sum of squares due to error after computing the total sum of squares and the sum of squares due to various factors. Derivation of Equation (3.9) as well as detailed mathematical description of ANOVA can be found in many books on statistics, such as Scheffé [S1], Rao [R3], and Searle [S2].

For the matrix experiment described in this chapter, Equation (3.9) implies:

(Total sum of squares) = (sum of the sums of squares

due to factors A, B, C, and D)

+ (sum of squares due to error) .

Note that the various sums of squares tabulated in Table 3.4 do satisfy the above equation.

Degrees of Freedom

The number of independent parameters associated with an entity like a matrix experiment, or a factor, or a sum of squares is called its *degrees of freedom*. A matrix experiment with nine rows has nine degrees of freedom and so does the grand total sum of squares. The overall mean has one degree of freedom and so does the sum of squares due to mean. Thus, the degrees of freedom associated with the total sum of squares is $9 - 1 = 8$. (Note that total sum of squares is equal to grand total sum of squares minus the sum of squares due to mean.)

Factor A has three levels, so its effect can be characterized by three parameters: a_1, a_2, and a_3. But these parameters must satisfy the constrain given by Equation (3.6). Thus, effectively, factor A has only two independent parameters and, hence, two degrees of freedom. Similarly, factors B, C, and D have two degrees of freedom each. In general, the degrees of freedom associated with a factor is one less than the number of levels.

The orthogonality of the matrix experiment implies the following relationship among the various degrees of freedom:

(Degrees of freedom for the total sum of squares)

= (sum of the degrees of freedom for the various factors)

+ (degrees of freedom for the error) .　　　　　　　(3.10)

Note the similarity between Equations (3.9) and (3.10). Equation (3.10) is useful for computing the degrees of freedom for error. In the present case study, the degrees of freedom for error comes out to be zero. This is consistent with the earlier observation that the error term is identically zero for each experiment in this case study.

It is customary to write the analysis of variance in a tabular form shown in Table 3.4. The mean square for a factor is computed by dividing the sum of squares by the degrees of freedom.

Estimation of Error Variance

The error variance, which is equal to the error mean square, can then be estimated as follows:

$$\text{Error variance} = \frac{\text{sum of squares due to error}}{\text{degrees of freedom for error}} . \qquad (3.11)$$

The error variance is denoted by σ_e^2.

In the interest of gaining the most information from a matrix experiment, all or most of the columns should be used to study process or product parameters. As a result, no degrees of freedom may be left to estimate error variance. Indeed, this is the situation with the present example. In such situations, we cannot directly estimate the error variance.

However, an approximate estimate of the error variance can be obtained by pooling the sum of squares corresponding to the factors having the lowest mean square. As a rule of thumb, we suggest that the sum of squares corresponding to the bottom half of the factors (as defined by lower mean square) corresponding to about half of the degrees of freedom be used to estimate the error mean square or error variance. This rule is similar to considering the bottom half harmonics in a Fourier expansion as error and using the rest to explain the function being investigated. In the present example, we use factors C and D to estimate the error mean square. Together, they account for four degrees of freedom and the sum of their sum of squares is 400. Hence, the error variance is 100. Error variance computed in this manner is indicated by parentheses, and the computation method is called *pooling*. (By the traditional statistical assumptions, pooling gives a biased estimate of error variance. To obtain a better estimate of error variance, a significantly larger number of experiments would be needed, the cost of which is usually not justifiable compared to the added benefit.)

In Fourier analysis of a signal, it is common to compute the power in all harmonics and then use only those harmonics with large power to explain the signal and treat the rest as error. Pooling of sum of squares due to bottom half factors is exactly analogous to that practice. After evaluating the sum of squares due to all factors, we retain only the top half factors to explain the variation in the process response η and the rest to estimate approximately the error variance.

The estimation of the error variance by pooling will be further illustrated through the applications discussed in the subsequent chapters. As it will be apparent from these applications, deciding which factors' sum of squares should be included in the error variance is usually obvious by inspecting the mean square column. The decision process can sometimes be improved by using a graphical data analysis technique called half-normal plots (see Daniel [D1] and Box, Hunter, and Hunter [B3]).

Confidence Intervals for Factor Effects

Confidence intervals for factor effects are useful in judging the size of the change caused by changing a factor level compared to the error standard deviation. As shown in Section 3.3, the variance of the effect of each factor level for this example is $(1/3)\sigma_e^2 = (1/3)(100) = 33.3$ $(dB)^2$. Thus, the width of the two-standard-deviation confidence interval, which is approximately 95 percent confidence interval, for each estimated effect is $\pm 2\sqrt{33.3} = \pm 11.5$ dB. In Figure 3.1 these confidence intervals are plotted for only the starting level to avoid crowding.

Variance Ratio

The variance ratio, denoted by F in Table 3.4, is the ratio of the mean square due to a factor and the error mean square. A large value of F means the effect of that factor is large compared to the error variance. Also, the larger the value of F, the more important that factor is in influencing the process response η. So, the values of F can be used to rank order the factors.

In statistical literature, the F value is often compared with the quantiles of a probability distribution called the F-distribution to determine the degree of confidence that a particular factor effect is real and not just a random occurrence (see, for example, Hogg and Craig [H3]). However, in Robust Design we are not concerned with such probability statements; we use the F ratio for only qualitative understanding of the relative factor effects. A value of F less than one means the factor effect is smaller than the error of the additive model. A value of F larger than two means the factor is not quite small, whereas larger than four means the factor effect is quite large.

Interpretation of ANOVA Tables

Thus far in this section, we have described the computation involved in the ANOVA table, as well as the inferences that can be made from the table. A variety of computer programs can be used to perform the calculations, but the experimenter must make appropriate inferences. Here we put together the major inferences from the ANOVA table.

Referring to the sum of squares column in Table 3.4, notice that factor A makes the largest contribution to the total sum of squares, namely, $(2450/3800) \times 100 = 64.5$ percent. Factor B makes the next largest contribution, $(950/3800) \times 100 = 25.0$ percent, to the total sum of squares. Factors C and D together make only a 10.5 percent contribution to the total sum of squares. The larger the contribution of a particular factor to the total sum of squares, the larger the ability is of that factor to influence η.

In this matrix experiment, we have used all the degrees of freedom for estimating the factor effects (four factors with two degrees of freedom each make up all the eight degrees of freedom for the total sum of squares). Thus, there are no degrees of freedom left for estimating the error variance. Following the rule of thumb spelled out earlier in this section, we use the bottom half factors that have the smallest mean square to estimate the error variance. Thus, we obtain the error sum of squares, indicated by parentheses in the ANOVA table, by pooling the sum of squares due to factors C and D. This gives 100 as an estimate of the error variance.

The largeness of a factor effect relative to the error variance can be judged from the F column. The larger the F value, the larger the factor effect is compared to the error variance.

This section points out that our purpose in conducting ANOVA is to determine the relative magnitude of the effect of each factor on the objective function η and to estimate the error variance. We do not attempt to make any probability statements about the significance of a factor as is commonly done in statistics. In Robust Design, ANOVA is also used to choose from among many alternatives the most appropriate quality characteristic and S/N ratio for a specific problem. Such an application of ANOVA is described in Chapter 8. Also, ANOVA is useful in computing the S/N ratio for dynamic problems as described in Chapter 9.

3.5 PREDICTION AND DIAGNOSIS

Prediction of η under Optimum Conditions

As discussed earlier, a primary goal of conducting Robust Design experiments is to determine the optimum level for each factor. For the CVD project, one of the two identified optimum conditions is $A_1 B_1 C_2 D_2$. The additive model, Equation (3.5), can be used to predict the value of η under the optimum conditions, denoted by η_{opt}, as follows:

$$\eta_{\text{opt}} = m + (m_{A_1} - m) + (m_{B_1} - m)$$

$$= -41.67 + (-20 + 41.67) + (-30 + 41.67)$$

$$= -8.33 \text{ dB} . \tag{3.12}$$

Note that since the sum of squares due to factors C and D are small and that these terms are included as error, we do not include the corresponding improvements in the prediction of η under optimum conditions. Why are the contributions by factors having a small sum of squares ignored? Because if we include the contribution from all factors, it can be shown that the predicted improvement in η exceeds the actual realized improvement—that is, our prediction would be biased on the higher side. By ignoring the contribution from factors with small sums of squares, we can reduce this bias. Again, this is a rule of thumb. For more precise prediction, we need to use appropriate *shrinkage coefficients* described by Taguchi [T1].

Thus, by Equation (3.12) we predict that the defect count under the optimum conditions would be -8.33 dB. This is equivalent to a mean square count of

$$y = 10^{-\frac{\eta_{opt}}{10}}$$

$$= 10^{0.833} = 6.8 \text{ (defects/unit area)}^2 .$$

The corresponding root-mean-square defect count is $\sqrt{6.8} = 2.6$ defects/unit area.

The purpose of taking log in constructing the S/N ratio can be explained in terms of the additive model. If the actual defect count were used as the characteristic for constructing the additive model, it is quite possible that the defect count predicted under the optimum conditions would have been negative. This is highly undesirable since negative counts are meaningless. However, in the log scale, such negative counts cannot occur. Hence, it is preferable to take the log.

The additive model is also useful in predicting the difference in defect counts between two process conditions. The anticipated improvement in changing the process conditions from the initial settings ($A_2 B_2 C_1 D_1$) to the optimum settings ($A_1 B_1 C_2 D_2$) is

$$\Delta\eta = \eta_{opt} - \eta_{initial} = (m_{A_1} - m_{A_2}) + (m_{B_1} - m_{B_2})$$

$$= (-20+45) + (-30+40)$$

$$= 35 \text{ dB} . \tag{3.13}$$

Once again we do not include the terms corresponding to factors C and D for the reasons explained earlier.

Verification (Confirmation) Experiment

After determining the optimum conditions and predicting the response under these conditions, we conduct an experiment with optimum parameter settings and compare the observed value of η with the prediction. If the predicted and observed η are close to each other, then we may conclude that the additive model is adequate for describing the dependence of η on the various parameters. On the contrary, if the observation is drastically different from the prediction, then we say the additive model is inadequate. This is evidence of a strong interaction among the parameters, which is described later in this section.

Variance of Prediction Error

We need to determine the variance of the prediction error so that we can judge the closeness of the observed η_{opt} to the predicted η_{opt}. The prediction error, which is the difference between the observed η_{opt} and the predicted η_{opt}, has two independent components. The first component is the error in the prediction of η_{opt} caused by the errors in the estimates of m, m_{A_1}, and m_{B_1}. The second component is the repetition error of an experiment. Because these two components are independent, the variance of the prediction error is the sum of their respective variances.

Consider the first component. Its variance can be shown equal to $(1/n_0)\sigma_e^2$ where σ_e^2 is the error variance whose estimation was discussed earlier; n_0 is the equivalent sample size for the estimation of η_{opt}. The equivalent sample size n_0 can be computed as follows:

$$\frac{1}{n_0} = \frac{1}{n} + \left[\frac{1}{n_{A_1}} - \frac{1}{n} \right] + \left[\frac{1}{n_{B_1}} - \frac{1}{n} \right] \tag{3.14}$$

where n is the number of rows in the matrix experiment and n_{A_1} is the number of times level A_1 was repeated in the matrix experiment—that is, n_{A_1} is the replication number for factor level A_1 and n_{B_1} is the replication number for factor level B_1.

Observe the correspondence between Equations (3.14) and (3.12). The term $(1/n)$ in Equation (3.14) corresponds to the term m in the prediction Equation (3.12); and the terms $(1/n_{A_1} - 1/n)$ and $(1/n_{B_1} - 1/n)$ correspond, respectively, to the terms $(m_{A_1} - m)$ and $(m_{B_1} - m)$. This correspondence can be used to generalize Equation (3.14) to other prediction formulae.

Now, consider the second component. Suppose we repeat the verification experiment n_r times under the optimum conditions and call the average η for these experiments as the observed η_{opt}. The repetition error is given by $(1/n_r)\sigma_e^2$. Thus, the variance of the prediction error, σ_{pred}^2, is

$$\sigma_{pred}^2 = \left[\frac{1}{n_0} \right] \sigma_e^2 + \left[\frac{1}{n_r} \right] \sigma_e^2 . \tag{3.15}$$

In the example, $n = 9$ and $n_{A_1} = n_{B_1} = 3$. Thus, $(1/n_0) = (1/9) + (1/3 - 1/9) + (1/3 - 1/9) = (5/9)$. Suppose $n_r = 4$. Then

$$\sigma_{pred}^2 = \left[\frac{5}{9} \right] \sigma_e^2 + \left[\frac{1}{4} \right] \sigma_e^2 = 80.6 (\text{dB})^2 .$$

The corresponding two-standard-deviation confidence limits for the prediction error are ±17.96 dB. If the prediction error is outside these limits, we should suspect the possibility that the additive model is not adequate. Otherwise, we consider the additive model to be adequate.

Uniformity of Prediction Error Variance

It is obvious from Equation (3.15) that the variance of the prediction error, σ^2_{pred}, is the same for all combinations of the factor levels in the experimental region. It does not matter whether the particular combination does or does not correspond to one of the rows in the matrix experiment. Before conducting the matrix experiment we do not know what would be the optimum combination. Hence, it is important to have the property of uniform prediction error.

Interactions among Control Factors

The concept of interactions can be understood from Figure 3.3. Figure 3.3(a) shows the case of no interaction between two factors A and B. Here, the lines of the effect of factor A for the settings B_1, B_2, and B_3 of factor B are parallel to each other. Parallel lines imply that if we change the level of factor A from A_1 to A_2 or A_3, the corresponding change in η is the same regardless of the level of factor B. Similarly, a change in level of B produces the same change in η regardless of the level of factor A. The additive model is perfect for this situation. Figures 3.3(b) and 3.3(c) show two examples of presence of interaction. In Figure 3.3(b), the lines are not parallel, but the direction of improvement does not change. In this case, the optimum levels identified by the additive model are still valid. Whereas in Figure 3.3(c), not only are the lines not parallel, but the direction of improvement is also not consistent. In such a case, the optimum levels identified by the additive model can be misleading. The type of interaction in Figure 3.3(b) is sometimes called *synergistic interaction* while the one in Figure 3.3(c) is called *antisynergistic interaction*. The concept of interaction between two factors described above can be generalized to apply to interaction among three or more factors.

When interactions between two or more factors are present, we need cross product terms to describe the variation of η in terms of the control factors. A model for such a situation needs more parameters than an additive model and, hence, it needs more experiments to estimate all the parameters. Further, as discussed in Chapter 6, using a model with interactions can have problems in the field. Thus, we consider the presence of interactions to be highly undesirable and try to eliminate them.

When the quality characteristic is correctly chosen, the S/N ratio is properly constructed, and the control factors are judiciously chosen (see Chapter 6 for guidelines), the additive model provides excellent approximation for the relationship between η and the control factors. *The primary purpose of the verification experiment is to warn us*

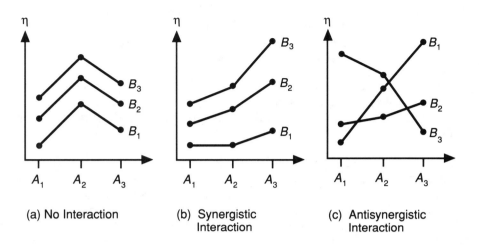

(a) No Interaction   (b) Synergistic Interaction   (c) Antisynergistic Interaction

Figure 3.3   Examples of interaction.

when the additive model is not adequate and, thus, prevent faulty process and product designs from going downstream. Some applications call for a broader assurance of the additive model. In such cases, the verification experiment consists of two or more conditions rather than just the optimum conditions. For the additive model to be considered adequate, the predictions must match the observation under all conditions that are tested. Also, in certain situations, we can judge from engineering knowledge that particular interactions are likely to be important. Then, orthogonal arrays can be suitably constructed to estimate those interactions along with the main effects, as described in Chapter 7.

3.6 SUMMARY

- A *matrix experiment* consists of a set of experiments where the settings of several product or process parameters to be studied are changed from one experiment to another.

- Matrix experiments are also called *designed experiments*, parameters are also called *factors*, and parameter settings are also called *levels*.

- Conducting matrix experiments using orthogonal arrays is an important technique in Robust Design. It gives more reliable estimates of factor effects with fewer experiments when compared to the traditional methods, such as one factor at a time experiments. Consequently, more factors can be studied in given R&D resources, leading to more robust and less expensive products.

- The columns of an orthogonal array are pairwise orthogonal—that is, for every pair of columns, all combinations of factor levels occur an equal number of times. The columns of the orthogonal array represent factors to be studied and the rows represent individual experiments.

- Conducting a matrix experiment with an orthogonal array is analogous to finding the frequency response function of a dynamic system by using a multifrequency input. The analysis of data obtained from matrix experiments is analogous to Fourier analysis.

- Some important terms used in matrix experiments are: The region formed by the factors being studied and their alternate levels is called the *experimental region*. The *starting levels* of the factors are the levels used before conducting the matrix experiment. The *main effects* of the factors are their separate effects. If the effect of a factor depends on the level of another factor, then the two factors are said to have an *interaction*. Otherwise, they are considered to have no interaction. The *replication number* of a factor level is the number of experiments in the matrix experiment that are conducted at that factor level. The *effect of a factor level* is the deviation it causes from the overall mean response. The *optimum level of a factor* is the level that gives the highest S/N ratio.

- An *additive model* (also called *superposition model* or *variables separable model*) is used to approximate the relationship between the response variable and the factor levels. Interactions are considered errors in the additive model.

- Orthogonal array based matrix experiments are used for a variety of purposes in Robust Design. They are used to:

  — Study the effects of control factors

  — Study the effects of noise factors

  — Evaluate the S/N ratio

  — Determine the best quality characteristic or S/N ratio for particular applications

- Key steps in analyzing data obtained from a matrix experiment are:

  1. Compute the appropriate summary statistics, such as the S/N ratio for each experiment.

  2. Compute the main effects of the factors.

  3. Perform ANOVA to evaluate the relative importance of the factors and the error variance.

  4. Determine the optimum level for each factor and predict the S/N ratio for the optimum combination.

5. Compare the results of the verification experiment with the prediction. If the results match the prediction, then the optimum conditions are considered confirmed; otherwise, additional analysis and experimentation are needed.

- If one or more experiments in a matrix experiment are missing or erroneous, then those experiments should be repeated to complete the matrix. This avoids the need for complicated analysis.

- Matrix experiment, followed by a verification experiment, is a powerful tool for detecting the presence of interactions among the control factors. If the predicted response under the optimum conditions does not match the observed response, then it implies that the interactions are important. If the predicted response matches the observed response, then it implies that the interactions are probably not important and that the additive model is a good approximation.

Chapter 4

STEPS IN ROBUST DESIGN

As explained in Chapter 2, optimizing a product or process design means determining the best architecture, levels of control factors, and tolerances. Robust Design is a methodology for finding the optimum settings of the control factors to make the product or process insensitive to noise factors. It involves eight steps that can be grouped into the three major categories of planning experiments, conducting them, and analyzing and verifying the results.

- **Planning the experiment**

    1) Identify the main function, side effects, and failure modes.

    2) Identify noise factors and the testing conditions for evaluating the quality loss.

    3) Identify the quality characteristic to be observed and the objective function to be optimized.

    4) Identify the control factors and their alternate levels.

    5) Design the matrix experiment and define the data analysis procedure.

- **Performing the experiment**

    6) Conduct the matrix experiment.

- **Analyzing and verifying the experiment results**

   7)   Analyze the data, determine optimum levels for the control factors, and predict performance under these levels.

   8)   Conduct the verification (also called *confirmation*) experiment and plan future actions.

These eight steps make up a Robust Design cycle. We will illustrate them in this chapter by using a case study of improving a polysilicon deposition process. The case study was conducted by Peter Hey in 1984 as a class project for the first offering of the 3-day Robust Design course developed by the author, Madhav Phadke, and Chris Sherrerd, Paul Sherry, and Rajiv Keny of AT&T Bell Laboratories. Hey and Sherry jointly planned the experiment and analyzed the data. The experiment yielded a 4-fold reduction in the standard deviation of the thickness of the polysilicon layer and nearly two orders of magnitude reduction in surface defects, a major yield-limiting problem which was virtually eliminated. These results were achieved by studying the effects of six control factors by conducting experiments under 18 distinct combinations of the levels of these factors—a rather small investment for huge benefits in quality and yield.

This chapter consists of nine sections:

- Sections 4.1 through 4.8 describes in detail the polysilicon deposition process case study in terms of the eight steps that form a Robust Design cycle.

- Section 4.9 summarizes the important points of this chapter.

4.1 THE POLYSILICON DEPOSITION PROCESS AND ITS MAIN FUNCTION

Manufacturing very large scale intergrated (VLSI) circuits involves about 150 major steps. Deposition of polysilicon comes after about half of the steps are complete, and, as a result, the silicon wafers (thin disks of silicon) used in the process have a significant amount of value added by the time they reach this step. The polysilicon layer is very important for defining the gate electrodes for the transistors. There are over 250,000 transistors in a square centimeter chip area for the 1.75 micron (micrometer ≡ micron) design rules used in the case study.

A hot-wall, reduced-pressure reactor (see Figure 4.1) is used to deposit polysilicon on a wafer. The reactor consists of a quartz tube which is heated by a 3-zone furnace. Silane and nitrogen gases are introduced at one end and pumped out the other. The silane gas pyrolizes, and a polysilicon layer is deposited on top of the oxide layer on the wafers. The wafers are mounted on quartz carriers. Two carriers, each carrying 25 wafers, can be placed inside the reactor at a time so that polysilicon is deposited simultaneously on 50 wafers.

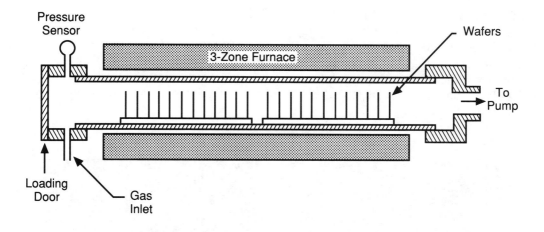

Figure 4.1   Schematic diagram of a reduced pressure reactor.

The function of the polysilicon deposition process is to deposit a uniform layer of a specified thickness. In the case study, the experimenters were interested in achieving 3600 angstrom(Å) thickness ($1\text{Å} = 10^{-10}$ meter). Figure 4.2 shows a cross section of the wafer after the deposition of the polysilicon layer.

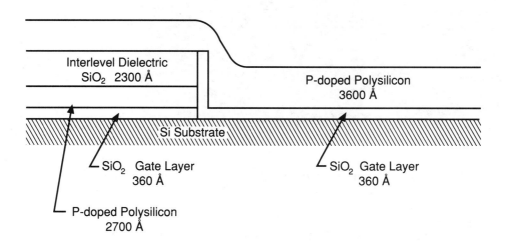

Figure 4.2   Cross section of a wafer showing polysilicon layer.

At the start of the study, two main problems occurred during the deposition process: (1) too many surface defects (see Figure 4.3) were encountered, and (2) too large a thickness variation existed within wafers and among wafers. In a subsequent VLSI manufacturing step, the polysilicon layer is patterned by an etching process to form lines of appropriate width and length. Presence of surface defects causes these lines to have variable width, which degrades the performance of the integrated circuits. The nonuniform thickness is detrimental to the etching process because it can lead to residual polysilicon in some areas and an etching away of the underlying oxide layer in other areas.

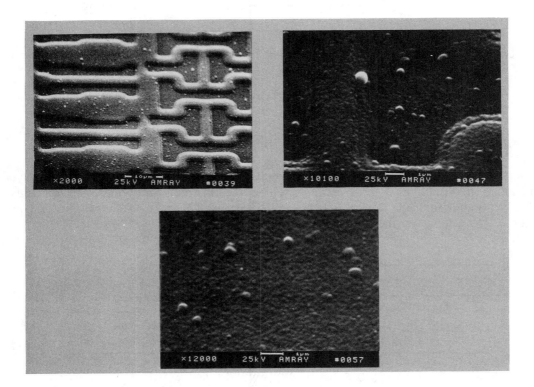

Figure 4.3   Photographs of polysilicon surface showing surface defects.

Prior to the case study, Hey noted that the surface-defect problem was crucial because a significant percentage of wafers were scrapped due to excessive defects. Also, he observed that controlling defect formation was particularly difficult due to its intermittent occurrence; for example, some batches of wafers (50 wafers make one batch) had approximately ten defects per unit area, while other batches had as many as 5,000 defects per unit area. Furthermore, no theoretical models existed to predict defect formation as a function of the various process parameters; therefore, experimentation was the only way to control the surface-defect problem. However, the intermittency of the problem had rendered the traditional method of experimentation, where only one process parameter is changed at a time, virtually useless.

4.2 NOISE FACTORS AND TESTING CONDITIONS

To minimize sensitivity to noise factors, we must first be able to estimate the sensitivity in a consistent manner for any combination of the control factor levels. This is achieved through proper selection of testing conditions.

In a Robust Design project, we identify all noise factors (factors whose levels cannot be controlled during manufacturing, which are difficult to control, or expensive to control), and then select a few testing conditions that capture the effect of the more important noise factors. Simulating the effects of all noise factors is impractical because the experimenter may not know all the noise sources and because total simulation would require too many testing conditions and be costly. Although it is not necessary to include the effect of all noise factors, the experimenter should list as many of them as possible and, then, use engineering judgment to decide which are more important and what testing conditions are appropriate to capture their effects.

Various noise factors exist in the deposition process. The nonuniform thickness and the surface defects of the polysilicon layer are caused by the variations in the parameters involved in the chemical reactions associated with the deposition process. First, the gases are introduced at one end of the reactor (see Figure 4.1). As they travel to the other end, the silane gas decomposes into polysilicon, which is deposited on the wafers, and into hydrogen. This activity causes a *concentration gradient* along the length of the reactor. Further, the *flow pattern* (direction and speed) of the gases need not be the same as they travel from one end of the tube to the other. The flow pattern could also vary from one part of a wafer to other parts of the same wafer. Another important noise factor is the *temperature* variation along the length and cross section of the tube. There are, of course, other sources of variation or noise factors, such as *topography of the wafer surface* before polysilicon deposition, variation in *pumping speed*, and variation in *gas supply*.

For the case study of the polysilicon deposition process, Hey and Sherry decided to process one batch of 50 wafers to evaluate the quality associated with each combination of control factor settings suggested by the orthogonal array experiment. Of these 50 wafers, only 3 were test wafers, while the remaining 47 were dummy wafers, which provided the needed "full load" effect while saving the cost of expensive test wafers. To capture the variation in reactant concentration, flow pattern variation, and temperature variation along the length of the tube, the test wafers were placed in positions 3, 23, and 48 along the tube. Furthermore, to capture the effect of noise variation across a wafer, the thickness and surface defects were measured at three points on each test wafer: top, middle, and bottom. Other noise factors were judged to be less important. To include their effect, the experimenters would have had to process multiple batches, thus making the experiments very expensive. Consequently, the other noise factors were ignored.

The testing conditions for this case study are rather simple: observe thickness and surface defects at three positions of three wafers, which are placed in specific positions along the length of the reactor. Sometimes orthogonal arrays (called *noise orthogonal arrays*) are used to determine the testing conditions that capture the effect of many noise factors. In some other situations, the technique of *compound noise factor* is used. These two techniques of constructing testing conditions are described in Chapter 8.

4.3 QUALITY CHARACTERISTICS AND OBJECTIVE FUNCTIONS

It is often tempting to observe the percentage of units that meet the specification and use that percentage directly as an objective function to be optimized. But, such temptation should be meticulously avoided. Besides being a poor measure of quality loss, using percentage of good (or bad) wafers as an objective function leads to orders of magnitude reduction in efficiency of experimentation. First, to observe accurately the percentage of "good" wafers, we need a large number (much larger than three) of test wafers for each combination of control factor settings. Secondly, when the percentage of good wafers is used as an objective function, the interactions among control factors often become dominant; consequently, additive models cannot be used as adequate approximations. The appropriate quality characteristics to be measured for the polysilicon deposition process in the case study were the polysilicon thickness and the surface defect count. The specifications were that the thickness should be within ± 8 percent of the target thickness and that the surface defect count should not exceed 10 per square centimeter.

As stated in Section 4.2, nine measurements (3 wafers × 3 measurements per wafer) of thickness and surface defects were taken for each combination of control factor settings in the matrix experiment. The ideal value for surface defects is zero—the smaller the number of surface defects per cm^2, the better the wafer. So, by adopting the quadratic loss function, we see that the objective function to be maximized is

$$\eta = -10 \log_{10} \text{ (mean square surface defects)}$$

$$= -10 \log_{10} \left\{ \frac{1}{9} \sum_{i=1}^{3} \sum_{j=1}^{3} y_{ij}^2 \right\} \tag{4.1}$$

where y_{ij} is the observed surface defect count at position j on test wafer i. Note that $j=1$, 2, and 3 stand for top, center, and bottom positions, respectively, on a test wafer. And $i=1$, 2, and 3 refer to position numbers 3, 23, and 48, respectively, along the length of the tube. Maximizing η leads to minimization of the quality loss due to surface defects.

The target value in the study for the thickness of the polysilicon layer was $\tau_0 = 3600 \,\overset{\circ}{A}$. Let τ_{ij} be the observed thickness at position j on test wafer i. The mean and variance of the thickness are given by

$$\mu = \frac{1}{9} \sum_{i=1}^{3} \sum_{j=1}^{3} \tau_{ij} \tag{4.2}$$

$$\sigma^2 = \frac{1}{8} \sum_{i=1}^{3} \sum_{j=1}^{3} (\tau_{ij} - \mu)^2. \tag{4.3}$$

The goal in optimization for thickness is to minimize variance while keeping the mean on target. This is a constrained optimization problem, which can be very difficult, especially when many control factors exist. However, as Chapter 5 shows, when a *scaling factor* (a factor that increases the thickness proportionally at all points on the wafers) exists, the problem can be simplified greatly.

In the case study, the deposition time was a clear scaling factor—that is, for every surface area where polysilicon was deposited, (thickness) = (deposition rate) × (deposition time). The deposition rate may vary from one wafer to the next, or from one position on a wafer to another position, due to the various noise factors cited in the previous section. However, the thickness at any point is proportional to the deposition time.

Thus, the constrained optimization problem in the case study can be solved in two steps as follows:

1. Maximize the Signal-to-noise (S/N) ratio, η',

$$\eta' = 10 \log_{10} \frac{\mu^2}{\sigma^2}. \tag{4.4}$$

2. Adjust the deposition time so that mean thickness is on target.

In summary, the two quality characteristics to be measured were the surface defects and the thickness. The corresponding objective functions to be maximized were η and η' defined by Equations (4.1) and (4.4), respectively. (Note that S/N ratio is a general term used for measuring sensitivity to noise factors. It takes a different form depending on the type of quality characteristic, as discussed in detail in Chapter 5. Both η and η' are different types of S/N ratios.)

The economics of a manufacturing process is determined by the throughput as well as by the quality of the products produced. Therefore, along with the quality characteristics, a throughput characteristic also must be studied. Thus, in the case study, the experimenters also observed the deposition rate, r, measured in angstroms of thickness growth per minute.

4.4 CONTROL FACTORS AND THEIR LEVELS

Processes, such as polysilicon deposition, typically have a large number of control factors (factors that can be freely specified by the process designer). The more complex a process, the more control factors it has and vice versa. Typically, we choose six to eight control factors at a time to optimize a process. For each factor we generally select two or three levels (or settings) and take the levels sufficiently far apart so that a wide region can be covered by the three levels. Commonly, one of these levels is taken to be the initial operating condition. Note that we are interested in the nonlinearity, so taking the levels of control factors too close together is not very fruitful. If we take only two levels, curvature effects would be missed, whereas such effects can be identified by selecting three levels for a factor (see Figure 4.4). Furthermore, by selecting three levels, we can simultaneously explore the region on either side of the initial operating condition. Hence, we prefer three levels.

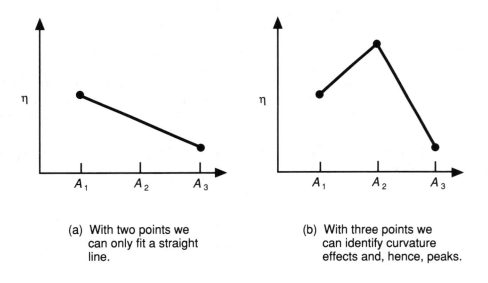

(a) With two points we can only fit a straight line.

(b) With three points we can identify curvature effects and, hence, peaks.

Figure 4.4   Linear and curvature effects of a factor.

In the case study, six control factors were selected for optimization. These factors and their alternate levels are listed in Table 4.1. The deposition temperature (A) is the steady state temperature at which the deposition takes place. When the wafers are placed in the reactor, they first have to be heated from room temperature to the deposition temperature and then held at that temperature. The deposition pressure (B) is the constant pressure maintained inside the reactor through appropriate pump speed and butterfly adjustment. The nitrogen flow (C) and the silane flow (D) are adjusted using the corresponding flow meters on gas tanks. Settling time (E) is the time between placing the wafer carriers in the reactors and the time at which gases flow. The settling time is important for establishing thermal and pressure equilibrium inside the reactor before the reaction is allowed to start. Cleaning method (F) refers to cleaning the wafers prior to the deposition step. Before undertaking the case study experiment, the practice was to perform no cleaning. The alternate two cleaning methods the experimenters wanted to study were CM_2, performed inside the reactor, and CM_3, performed outside the reactor.

TABLE 4.1 CONTROL FACTORS AND THEIR LEVELS

Factor	Levels*		
	1	2	3
A. Deposition temperature (°C)	$T_0 - 25$	$\underline{T_0}$	$T_0 + 25$
B. Deposition pressure (mtorr)	$P_0 - 200$	$\underline{P_0}$	$P_0 + 200$
C. Nitrogen flow (sccm)	$\underline{N_0}$	$N_0 - 150$	$N_0 - 75$
D. Silane flow (sccm)	$S_0 - 100$	$S_0 - 50$	$\underline{S_0}$
E. Settling time (min)	$\underline{t_0}$	$t_0 + 8$	$t_0 + 16$
F. Cleaning method	$\underline{None}$	CM_2	CM_3

* Starting levels are identified by underscore.

While deciding on the levels of control factors, a frequent tendency is to choose the levels relatively close to the starting levels. This is due to the experimenter's concern that a large number of bad products may be produced during the matrix experiment. But, producing bad products during the experiment stage may, in fact, be beneficial because it tells us which region of control factor levels should be avoided. Also, by choosing levels that are wide apart, we increase the chance of capturing the nonlinearity of the relationship between the control factors and the noise factors, and, thus, finding the levels of control factors that minimize sensitivity to noise. Further, when the levels are wide apart, the factor effects are large when compared to the experimental errors. As a result, the factor effects can be identified without too many repetitions.

Thus, it is important to resist the tendency to choose control factor levels that are rather close. Of course, during subsequent refinement experiments, levels closer to each other could be chosen. In the polysilicon deposition case study, the ratio of the largest to the smallest levels of factors B, C, D, and, E was between three and five which represents a wide variation. Temperature variation from $(T_0 - 25)$ °C to $(T_0 + 25)$ °C also represents a wide range in terms of the known impact on the deposition rate.

The initial settings of the six control factors are indicated by an *underscore* in Table 4.1. The objective of this project was to determine the optimum level for each factor so that η and η' are improved, while ensuring simultaneously that the deposition rate, r, remained as high as possible. Note that the six control factors and their selected settings define the experimental region over which process optimization was done.

4.5 MATRIX EXPERIMENT AND DATA ANALYSIS PLAN

An efficient way to study the effect of several control factors simultaneously is to plan matrix experiments using orthogonal arrays. As pointed out in Chapter 3, orthogonal arrays offer many benefits. First, the conclusions arrived at from such experiments are valid over the entire experimental region spanned by the control factors and their settings. Second, there is a large saving in the experimental effort. Third, the data analysis is very easy. Finally, it can detect departure from the additive model.

An orthogonal array for a particular Robust Design project can be constructed from the knowledge of the number of control factors, their levels, and the desire to study specific interactions. While constructing the orthogonal array, we also take into account the difficulties in changing the levels of control factors, other physical limitations in conducting experiments, and the availability of resources. In the polysilicon deposition case study, there were six factors, each at three levels. The experimenters found no particular reason to study specific interactions and no unusual difficulty in changing the levels of any factor. The available resources for conducting the experiments were such that about 20 batches could be processed and appropriate measurements made. Using the standard methods of constructing orthogonal arrays, which are described in Chapter 7, the standard array L_{18} was selected for this matrix experiment.

The L_{18} orthogonal array is given in Table 4.2. It has eight columns and eighteen rows. The first column is a 2-level column—that is, it has only two distinct entries, namely 1 or 2. All the chosen six control factors have three levels. So, column 1 was kept empty or unassigned. From the remaining seven 3-level columns, column 7 was arbitrarily designated as an empty column, and factors A through F were assigned, respectively, to columns 2 through 6 and 8. (Note that keeping one or more columns empty does not alter the orthogonality property of the array. Thus, the matrix formed by columns 2 through 6 and 8 is still an orthogonal array. But, if one or more rows are dropped, the orthogonality is destroyed.) The reader can verify the orthogonality by checking that for every pair of columns all combinations of levels occur, and they occur an equal number of times.

The 18 rows of the L_{18} array represent the 18 experiments to be conducted. Thus, experiment 1 is to be conducted at level 1 for each of the six control factors. These levels can be read from Table 4.1. However, to make it convenient for the experimenter and to prevent translation errors, the entire matrix of Table 4.2 should be translated using the level definitions in Table 4.1 to create the experimenter's log sheet shown in Table 4.3.

TABLE 4.2 L_{18} ORTHOGONAL ARRAY AND FACTOR ASSIGNMENT

Expt. No.	Column Numbers and Factor Assignment*							
	1 e	2 A	3 B	4 C	5 D	6 E	7 e	8 F
1	1	1	1	1	1	1	1	1
2	1	1	2	2	2	2	2	2
3	1	1	3	3	3	3	3	3
4	1	2	1	1	2	2	3	3
5	1	2	2	2	3	3	1	1
6	1	2	3	3	1	1	2	2
7	1	3	1	2	1	3	2	3
8	1	3	2	3	2	1	3	1
9	1	3	3	1	3	2	1	2
10	2	1	1	3	3	2	2	1
11	2	1	2	1	1	3	3	2
12	2	1	3	2	2	1	1	3
13	2	2	1	2	3	1	3	2
14	2	2	2	3	1	2	1	3
15	2	2	3	1	2	3	2	1
16	2	3	1	3	2	3	1	2
17	2	3	2	1	3	1	2	3
18	2	3	3	2	1	2	3	1

* Empty columns are identified by e.

TABLE 4.3 EXPERIMENTER'S LOG

Expt. No.	Temperature	Pressure	Nitrogen	Silane	Settling Time	Cleaning Method
1	$T_0 - 25$	$P_0 - 200$	N_0	$S_0 - 100$	t_0	None
2	$T_0 - 25$	P_0	$N_0 - 150$	$S_0 - 50$	$t_0 + 8$	CM_2
3	$T_0 - 25$	$P_0 + 200$	$N_0 - 75$	S_0	$t_0 + 16$	CM_3
4	T_0	$P_0 - 200$	N_0	$S_0 - 50$	$t_0 + 8$	CM_3
5	T_0	P_0	$N_0 - 150$	S_0	$t_0 + 16$	None
6	T_0	$P_0 + 200$	$N_0 - 75$	$S_0 - 100$	t_0	CM_2
7	$T_0 + 25$	$P_0 - 200$	$N_0 - 150$	$S_0 - 100$	$t_0 + 16$	CM_3
8	$T_0 + 25$	P_0	$N_0 - 75$	$S_0 - 50$	t_0	None
9	$T_0 + 25$	$P_0 + 200$	N_0	S_0	$t_0 + 8$	CM_2
10	$T_0 - 25$	$P_0 - 200$	$N_0 - 75$	S_0	$t_0 + 8$	None
11	$T_0 - 25$	P_0	N_0	$S_0 - 100$	$t_0 + 16$	CM_2
12	$T_0 - 25$	$P_0 + 200$	$N_0 - 150$	$S_0 - 50$	t_0	CM_3
13	T_0	$P_0 - 200$	$N_0 - 150$	S_0	t_0	CM_2
14	T_0	P_0	$N_0 - 75$	$S_0 - 100$	$t_0 + 8$	CM_3
15	T_0	$P_0 + 200$	N_0	$S_0 - 50$	$t_0 + 16$	None
16	$T_0 + 25$	$P_0 - 200$	$N_0 - 75$	$S_0 - 50$	$t_0 + 16$	CM_2
17	$T_0 + 25$	P_0	N_0	S_0	t_0	CM_3
18	$T_0 + 25$	$P_0 + 200$	$N_0 - 150$	$S_0 - 100$	$t_0 + 8$	None

Now we combine the experimenter's log sheet with the testing conditions described in Section 4.2 to create the following experimental procedure:

1. Conduct 18 experiments as specified by the 18 rows of Table 4.3.

2. For each experiment, process one batch, consisting of 47 dummy wafers and three test wafers. The test wafers should be placed in positions 3, 23, and 48.

3. For each experiment, compute to your best ability the deposition time needed to achieve the target thickness of 3600Å. Note that in the experiment the actual thickness may turn out to be much different from 3600Å. However, such data are perfectly useful for analysis. Thus, a particular experiment *need not* be redone by adjusting the deposition time to obtain 3600Å thickness.

4. For each experiment, measure the surface defects and thickness at three specific points (top, center, and bottom) on each test wafer. Follow standard laboratory practice to prepare data sheets with space for every observation to be recorded.

4.6 CONDUCTING THE MATRIX EXPERIMENT

From Table 4.3 it is apparent that, from one experiment to the next, levels of several control factors must be changed. This poses a considerable amount of difficulty to the experimenter. Meticulousness in correctly setting the levels of the various control factors is critical to the success of a Robust Design project. Let us clarify what we mean by meticulousness. Going from experiment 3 to experiment 4 we must change temperature from $(T_0 - 25)$ °C to T_0 °C, pressure from $(P_0 + 200)$ mtorr to $(P_0 - 200)$ mtorr, and so on. By meticulousness we mean ensuring that the temperature, pressure, and other dials are set to their proper levels. Failure to set the level of a factor correctly could destroy the valuable property of orthogonality. Consequently, conclusions from the experiment could be erroneous. However, if an inherent error in the equipment leads to an actual temperature of $(T_0 - 1)$ °C or $(T_0 + 2)$ °C when the dial is set at T_0 °C, we should not bother to correct for such variations. Why? Because unless we plan to change the equipment, such variations constitute noise and will continue to be present during manufacturing. If our conclusions from the matrix experiment are to be valid in actual manufacturing, our results must not be sensitive to such inherent variations. By keeping these variations out of our experiments, we lose the ability to test for robustness against such variations. The matrix experiment, coupled with the verification experiment, has a built-in check for sensitivity to such inherent variations.

A difficulty in conducting matrix experiments is their radical difference from the current practice of conducting product or process design experiments. One common practice is to guess, using engineering judgment, the improved settings of the control factors and then conduct a paired comparison with the starting conditions. The guess-and-test cycle is repeated until some minimum improvement is obtained, the deadline is reached, or the budget is exhausted. This practice relies heavily on luck, and it is inefficient and time-consuming.

Another common practice is to optimize systematically one control factor at a time. Suppose we wish to determine the effect of the three temperature settings while keeping the settings of the other control factors fixed at their starting levels. To reduce the effect of experimental error, we must process several batches at each temperature setting. Suppose six batches are processed at each temperature setting. (Note that in the L_{18} array the replication number is six; that is, there are six experiments for each factor level.) Then, we would need 18 batches to evaluate the effect of three temperature settings. For the other factors, we need to experiment with the two alternate levels, so that we need to process 12 batches each. Thus, for the six factors, we would need to process $18 + 5 \times 12 = 78$ batches. This is a large number compared to the 18 batches needed for the matrix experiment. Further, if there are strong interactions among the control factors, this method of experimentation cannot detect them.

The matrix experiment, though somewhat tedious to conduct, is highly efficient—that is, when compared to the practices above, we can generate more dependable information about more control factors with the same experimental effort. Also, this method of experimentation allows for the detection of the interactions among the control factors, when they are present, through the verification experiment.

In practice, many design improvement experiments, where only one factor is studied at a time, get terminated after studying only a few control factors because both the R&D budget and the experimenter's patience run out. As a result, the quality improvement turns out to be only partial, and the product cost remains somewhat high. This danger is reduced greatly when we conduct matrix experiments using orthogonal arrays.

In the polysilicon deposition case study, the 18 experiments were conducted according to the experimenter's log given in Table 4.3. It took only nine days (2 experiments per day) to conduct them. The observed data on surface defects are listed in Table 4.4(a), and the thickness and deposition rate data are shown in Table 4.4(b). The surface defects were measured by placing the specimen under an optical microscope and counting the defects in a field of 0.2 cm^2. When the count was high, the field area was divided into smaller areas, defects in one area were counted, and the count was then multiplied by an appropriate number to determine the defect count per unit area (0.2 cm^2). The thickness was measured by an optical interferometer. The deposition rate was computed by dividing the average thickness by the deposition time.

4.7 DATA ANALYSIS

The first step in data analysis is to summarize the data for each experiment. For the case study, these calculations are illustrated next.

For experiment number 1, the S/N ratio for the surface defects, given by Equation (4.1), was computed as follows:

$$\eta = -10 \log_{10} \left[\frac{1}{9} \sum_{i=1}^{3} \sum_{j=1}^{3} y_{ij}^2 \right]$$

$$= -10 \log_{10} \left[\frac{(1^2 + 0^2 + 1^2) + (2^2 + 0^2 + 0^2) + (1^2 + 1^2 + 0^2)}{9} \right]$$

$$= -10 \log_{10} \left[\frac{8}{9} \right]$$

$$= 0.51.$$

From the thickness data, the mean, variance, and S/N ratio were calculated as follows by using Equations (4.2), (4.3) and (4.4):

See Eq. (4.2) $\mu = \dfrac{1}{9} \sum\limits_{i=1}^{3} \sum\limits_{j=1}^{3} \tau_{ij}$

$$= \frac{1}{9} \left[(2029 + 1975 + 1961) + (1975 + 1934 + 1907) + (1952 + 1941 + 1949) \right]$$

$$= 1958.1 \ \text{\AA} \ .$$

See Eq. (4.3) $\sigma^2 = \dfrac{1}{8} \sum\limits_{i=1}^{3} \sum\limits_{j=1}^{3} (\tau_{ij} - \mu)^2$

$$= \frac{1}{8} \left[(2029 - 1958.1)^2 + \cdots + (1949 - 1958.1)^2 \right]$$

$$= 1151.36 \ (\text{\AA})^2.$$

$$\eta' = 10 \log_{10} \frac{\mu^2}{\sigma^2}$$

$$= 10 \log_{10} \frac{1958.1^2}{1151.36}$$

$$= 35.22 \ \text{dB}.$$

TABLE 4.4(a) SURFACE DEFECT DATA (DEFECTS / UNIT AREA)

Expt. No.	Test Wafer 1			Test Wafer 2			Test Wafer 3		
	Top	Center	Bottom	Top	Center	Bottom	Top	Center	Bottom
1	1	0	1	2	0	0	1	1	0
2	1	2	8	180	5	0	126	3	1
3	3	35	106	360	38	135	315	50	180
4	6	15	6	17	20	16	15	40	18
5	1720	1980	2000	487	810	400	2020	360	13
6	135	360	1620	2430	207	2	2500	270	35
7	360	810	1215	1620	117	30	1800	720	315
8	270	2730	5000	360	1	2	9999	225	1
9	5000	1000	1000	3000	1000	1000	3000	2800	2000
10	3	0	0	3	0	0	1	0	1
11	1	0	1	5	0	0	1	0	1
12	3	1620	90	216	5	4	270	8	3
13	1	25	270	810	16	1	225	3	0
14	3	21	162	90	6	1	63	15	39
15	450	1200	1800	2530	2080	2080	1890	180	25
16	5	6	40	54	0	8	14	1	1
17	1200	3500	3500	1000	3	1	9999	600	8
18	8000	2500	3500	5000	1000	1000	5000	2000	2000

TABLE 4.4(b) THICKNESS AND DEPOSITION RATE DATA

Expt. No.	Thickness (Å)									Deposition Rate (Å/min)
	Test Wafer 1			Test Wafer 2			Test Wafer 3			
	Top	Center	Bottom	Top	Center	Bottom	Top	Center	Bottom	
1	2029	1975	1961	1975	1934	1907	1952	1941	1949	14.5
2	5375	5191	5242	5201	5254	5309	5323	5307	5091	36.6
3	5989	5894	5874	6152	5910	5886	6077	5943	5962	41.4
4	2118	2109	2099	2140	2125	2108	2149	2130	2111	36.1
5	4102	4152	4174	4556	4504	4560	5031	5040	5032	73.0
6	3022	2932	2913	2833	2837	2828	2934	2875	2841	49.5
7	3030	3042	3028	3486	3333	3389	3709	3671	3687	76.6
8	4707	4472	4336	4407	4156	4094	5073	4898	4599	105.4
9	3859	3822	3850	3871	3922	3904	4110	4067	4110	115.0
10	3227	3205	3242	3468	3450	3420	3599	3591	3535	24.8
11	2521	2499	2499	2576	2537	2512	2551	2552	2570	20.0
12	5921	5766	5844	5780	5695	5814	5691	5777	5743	39.0
13	2792	2752	2716	2684	2635	2606	2765	2786	2773	53.1
14	2863	2835	2859	2829	2864	2839	2891	2844	2841	45.7
15	3218	3149	3124	3261	3205	3223	3241	3189	3197	54.8
16	3020	3008	3016	3072	3151	3139	3235	3162	3140	76.8
17	4277	4150	3992	3888	3681	3572	4593	4298	4219	105.3
18	3125	3119	3127	3567	3563	3520	4120	4088	4138	91.4

The deposition rate in the decibel scale for experiment 1 is given by

$$\eta'' = 10 \log_{10} r^2 = 20 \log_{10} r$$

$$= 20 \log_{10}(14.5)$$

$$= 23.23 \text{ dBam}$$

where dBam stands for decibel $\overset{\circ}{A}$ /min.

The data summary for all 18 experiments was computed in a similar fashion and the results are tabulated in Table 4.5.

Observe that the mean thickness for the 18 experiments ranges from 1958 $\overset{\circ}{A}$ to 5965 $\overset{\circ}{A}$. But we are least concerned about this variation in the thickness because the average thickness can be adjusted easily by changing the deposition time. During a Robust Design project, what we are most interested in is the S/N ratio, which in this case is a measure of variation in thickness as a proportion of the mean thickness. Hence, no further analysis on the mean thickness was done in the case study, but the mean thickness, of course, was used in computing the deposition rate, which was of interest.

After the data for each experiment are summarized, the next step in data analysis is to estimate the effect of each control factor on each of the three characteristics of interest and to perform analysis of variance (ANOVA) as described in Chapter 3.

The factor effects for surface defects (η), thickness (η'), and deposition rate (η''), and the respective ANOVA are given in Tables 4.6, 4.7, and 4.8, respectively. A summary of the factor effects is tabulated in Table 4.9, and the factor effects are displayed graphically in Figure 4.5, which makes it easy to visualize the relative effects of the various factors on all three characteristics.

To assist the interpretation of the factor effects plotted in Figure 4.5, we note the following relationship between the decibel scale and the natural scale for the three characteristics:

- An increase in η by 6 dB is equivalent to a reduction in the root mean square surface defects by a factor of 2. An increase in η by 20 dB is equivalent to a reduction in the root mean square surface defects by a factor of 10.

- The above statements are valid if we substitute η' or η'' for η, and standard deviation of thickness or deposition rate for root mean square surface defects.

The task of determining the best setting for each control factor can become complicated when there are multiple characteristics to be optimized. This is because different levels of the same factor could be optimum for different characteristics. The quality loss function could be used to make the necessary trade-offs when different characteristics suggest different optimum levels. For the polysilicon deposition case

TABLE 4.5 DATA SUMMARY BY EXPERIMENT

Expt. No.	Experiment Condition Matrix* e A B C D E e F	Surface Defects η (dB)	Thickness μ ($\mathring{A}$)	Thickness η' (dB)	Deposition Rate η'' (dBam)
1	1 1 1 1 1 1 1 1	0.51	1958	35.22	23.23
2	1 1 2 2 2 2 2 2	−37.30	5255	35.76	31.27
3	1 1 3 3 3 3 3 3	−45.17	5965	36.02	32.34
4	1 2 1 1 2 2 3 3	−25.76	2121	42.25	31.15
5	1 2 2 2 3 3 1 1	−62.54	4572	21.43	37.27
6	1 2 3 3 1 1 2 2	−62.23	2891	32.91	33.89
7	1 3 1 2 1 3 2 3	−59.88	3375	21.39	37.68
8	1 3 2 3 2 1 3 1	−71.69	4527	22.84	40.46
9	1 3 3 1 3 2 1 2	−68.15	3946	30.60	41.21
10	2 1 1 3 3 2 2 1	−3.47	3415	26.85	27.89
11	2 1 2 1 1 3 3 2	−5.08	2535	38.80	26.02
12	2 1 3 2 2 1 1 3	−54.85	5781	38.06	31.82
13	2 2 1 2 3 1 3 2	−49.38	2723	32.07	34.50
14	2 2 2 3 1 2 1 3	−36.54	2852	43.34	33.20
15	2 2 3 1 2 3 2 1	−64.18	3201	37.44	34.76
16	2 3 1 3 2 3 1 2	−27.31	3105	31.86	37.71
17	2 3 2 1 3 1 2 3	−71.51	4074	22.01	40.45
18	2 3 3 2 1 2 3 1	−72.00	3596	18.42	39.22

* Empty column is denoted by e.

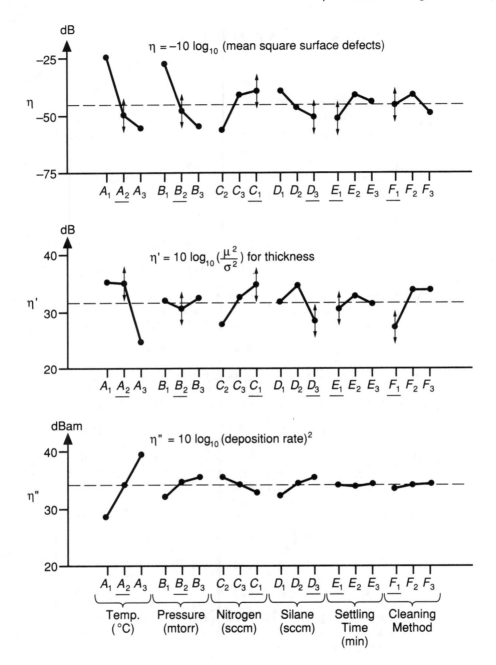

Figure 4.5 Plots of factor effects. Underline indicates starting level. Two-standard-deviation confidence limits are also shown for the starting level. Estimated confidence limits for η'' are too small to show.

study, we can make the following observations about the optimum setting from Figure 4.5 and Table 4.9:

- Deposition temperature (factor A) has the largest effect on all three characteristics. By reducing the temperature from the starting setting of T_0 °C to $T_0 - 25$ °C, η can be improved by $\{(-24.23) - (-50.10)\} \simeq 26$ dB. This is equivalent to a 20-fold reduction in root mean square surface defect count. The effect of this temperature change on thickness uniformity is only $(35.12 - 34.91) = 0.21$ dB, which is negligible. But the same temperature change would lead to a reduction in deposition rate by $(34.13 - 28.76) \simeq 5.4$ dB, which is approximately a 2-fold reduction in the deposition rate. Thus, temperature can dramatically reduce the surface defect problem, but it also would double the deposition time. Accordingly, there is a trade-off to be made between reducing the quality cost (including the scrap due to high surface defect count) and the number of wafers processed per day by the reactor.

- Deposition pressure (factor B) has the next largest effect on surface defect and deposition rate. Reducing the pressure from the starting level of P_0 mtorr to $(P_0 - 200)$ mtorr can improve η by about 20 dB (a 10-fold reduction in the root mean square surface defect count) at the expense of reducing the deposition rate by 2.75 dBam (37 percent reduction in deposition rate). The effect of pressure on thickness uniformity is very small.

- Nitrogen flow rate (factor C) has a moderate effect on all three characteristics. The starting setting of N_0 sccm gives the highest S/N ratios for surface defects and thickness uniformity. There is also a possibility of further improving these two S/N ratios by increasing the flow rate of this dilutant gas. This is an important fact to be remembered for future experiments. The effect of nitrogen flow rate on deposition rate is small compared to the effects of temperature and pressure.

- Silane flow rate (factor D) also has a moderate effect on all three characteristics. Thickness uniformity is the best when silane flow rate is set at $(S_0 - 50)$ sccm. This can also lead to a small reduction in surface defects and the deposition rate.

- Settling time (factor E) can be used to achieve about 10 dB improvement in surface defects by increasing the time from t_0 minutes to $(t_0 + 8)$ minutes. The data indicates that a further increase in the settling time to $(t_0 + 16)$ minutes could negate some of the reduction in surface defect count. However, this change is small compared to the standard deviation of the error; and it is not physically justifiable. Settling time has no effect on the deposition rate and the thickness uniformity.

- Cleaning method (factor F) has no effect on deposition rate and surface defects. But, by instituting some cleaning prior to deposition, the thickness uniformity can be improved by over 6.0 dB (a factor of 2 reduction in standard deviation of thickness). Cleaning with CM_2 or CM_3 could give the same improvement in thickness uniformity. However, CM_2 cleaning can be performed inside the reactor, whereas CM_3 cleaning must be done outside the reactor. Thus, CM_2 cleaning is more convenient.

From these observations, the optimum settings of factors E and F are obvious, namely E_2 and F_2. However, for factors A through D, the direction in which the quality characteristics (surface defects and thickness uniformity) improve tend to reduce the deposition rate. Thus, a trade-off between quality loss and productivity must be made in choosing their optimum levels. In the case study, since surface defects were the key quality problem that caused significant scrap, the experimenters decided to take care of it by changing temperature from A_2 to A_1. As discussed earlier, this also meant a substantial reduction in deposition rate. Also, they decided to hold the other three factors at their starting levels, namely $B_2,$ C_1, and D_3. The potential these factors held would

TABLE 4.6 ANALYSIS OF SURFACE DEFECTS DATA*

Factor	Average η by Factor Level (dB)			Degree of Freedom	Sum of Squares	Mean Square	F
	1	2	3				
A. Temperature	− 24.23	−50.10	− 61.76	2	4427	2214	27
B. Pressure	− 27.55	−47.44	− 61.10	2	3416	1708	21
C. Nitrogen	−39.03	− 55.99	− 41.07	2	1030	515	6.4
D. Silane	− 39.20	− 46.85	−50.04	2	372	186	2.3
E. Settling time	−51.52	− 40.54	− 44.03	2	378	189	2.3
F. Cleaning method	−45.56	− 41.58	− 48.95	2	164†	82	
Error				5	405†	81	
Total				17	10192		
(Error)				(7)	(569)	(81)	

* Overall mean η = −45.36 dB. Underscore indicates starting level.
† Indicates the sum of squares added together to form the pooled error sum of squares shown in parentheses.

have been used if the confirmation experiment indicated a need to improve the surface defect and thickness uniformity further. Thus, the optimum conditions chosen were: $A_1 B_2 C_1 D_3 E_2 F_2$.

The next step in data analysis is to predict the anticipated improvements under the chosen optimum conditions. To do so, we first predict the S/N ratios for surface defects, thickness uniformity, and deposition rate using the additive model. These computations for the case study are displayed in Table 4.10. According to the table, an improvement in surface defects equal to $[-19.84 - (-56.69)] = 36.85$ dB should be anticipated, which is equivalent to a reduction in the root mean square surface defect count by a factor of 69.6. The projected improvement in thickness uniformity is $36.79 - 29.95 = 6.84$ dB, which implies a reduction in standard deviation by a factor of 2.2. The corresponding change in deposition rate is $29.60 - 34.97 = -5.37$ dB, which amounts to a reduction in the deposition rate by a factor of 1.9.

TABLE 4.7 ANALYSIS OF THICKNESS DATA*

Factor	Average η' by Level (dB)			Degree of Freedom	Sum of Squares	Mean Square	F
	1	2	3				
A. Temperature	35.12	34.91	24.52	2	440	220	16
B. Pressure	31.61	30.70	32.24	2	7†	3.5	
C. Nitrogen	34.39	27.86	32.30	2	134	67	5.0
D. Silane	31.68	34.70	28.17	2	128	64	4.8
E. Settling time	30.52	32.87	31.16	2	18†	9	
F. Cleaning method	27.04	33.67	33.85	2	181	90.5	6.8
Error				5	96†	19.2	
Total				17	1004	59.1	
(Error)				(9)	(121)	(13.4)	

* Overall mean $\eta' = 31.52$ dB. Underscore indicates starting level.
† Indicates the sum of squares added together to form the pooled error sum of squares shown in parentheses.

TABLE 4.8 ANALYSIS OF DEPOSITION RATE DATA*

Factor	Average η'' by Factor Level (dBam)			Degree of Freedom	Sum of Squares	Mean Square	F
	1	2	3				
A. Temperature	28.76	34.13	39.46	2	343.1	171.5	553
B. Pressure	32.03	34.78	35.54	2	41.0	20.5	66
C. Nitrogen	32.81	35.29	34.25	2	18.7	9.4	30
D. Silane	32.21	34.53	35.61	2	36.3	18.1	58
E. Settling time	34.06	33.99	34.30	2	0.3†	0.2	
F. Cleaning method	33.81	34.10	34.44	2	1.2†	0.6	
Error				5	1.3†	0.26	
Total				17	441.9	25.9	
(Error)				(9)	(2.8)	(0.31)	

* Overall mean $\eta'' = 34.12$ dBam. Underscore indicates starting level.
† Indicates the sum of squares added together to form the pooled error sum of squares shown in
 parentheses.

4.8 VERIFICATION EXPERIMENT AND FUTURE PLAN

Conducting a verification experiment is a crucial final step of a Robust Design project.
Its purpose is to verify that the optimum conditions suggested by the matrix experiment
do indeed give the projected improvement. If the observed S/N ratios under the
optimum conditions are close to their respective predictions, then we conclude that the
additive model on which the matrix experiment was based is a good approximation of
the reality. Then, we adopt the recommended optimum conditions for our process or
product, as the case may be. However, if the observed S/N ratios under the optimum
conditions differ drastically from their respective predictions, there is an evidence of
failure of the additive model. There can be many reasons for the failure and, thus,
there are many ways of dealing with it. The failure of the additive model generally
indicates that choice of the objective function or the S/N ratio is inappropriate, the
observed quality characteristic was chosen incorrectly, or the levels of the control factors were chosen inappropriately. The question of how to avoid serious additivity
problems by properly choosing the quality characteristic, the S/N ratio, and the control
factors and their levels is discussed in Chapter 6. Of course, another way to handle the

TABLE 4.9 SUMMARY OF FACTOR EFFECTS

Factor	Level	Surface Defects		Thickness		Deposition Rate	
		η (dB)	F	η' (dB)	F	η'' (dBam)	F
A. Temperature (°C)	$A_1: T_0 - 25$	−24.23		35.12		28.76	
	$A_2: T_0$	−50.10	27	34.91	16	34.13	553
	$A_3: T_0 + 25$	−61.76		24.52		39.46	
B. Pressure (mtorr)	$B_1: P_0 - 200$	−27.55		31.61		32.03	
	$B_2: P_0$	−47.44	21	30.70	−	34.78	66
	$B_3: P_0 + 200$	−61.10		32.24		35.54	
C. Nitrogen (sccm)	$C_1: N_0$	−39.03		34.39		32.81	
	$C_2: N_0 - 150$	−55.99	6.4	27.86	5.0	35.29	30
	$C_3: N_0 - 75$	−41.07		32.30		34.25	
D. Silane (sccm)	$D_1: S_0 - 100$	−39.20		31.68		32.21	
	$D_2: S_0 - 50$	−46.85	2.3	34.70	4.8	34.53	58
	$D_3: S_0$	−50.04		28.17		35.61	
E. Settling time (min)	$E_1: t_0$	−51.52		30.52		34.06	
	$E_2: t_0 + 8$	−40.54	2.3	32.87	−	33.99	−
	$E_3: t_0 + 16$	−44.03		31.16		34.30	
F. Cleaning method	$F_1: None$	−45.56		27.04		33.81	
	$F_2: CM_2$	−41.58	−	33.67	6.8	34.10	−
	$F_3: CM_3$	−48.95		33.85		34.44	
Overall mean		−45.36		31.52		34.12	

additivity problem is to study a few key interactions among the control factors in future experiments. Construction of orthogonal arrays that permit the estimation of a few specific interactions, along with all main effects, is discussed in Chapter 7.

The verification experiment has two aspects: the first is that the predictions must agree under the laboratory conditions; the second aspect is that the predictions should be valid under actual manufacturing conditions for the process design and under actual field conditions for the product design. A judicious choice of both the noise factors to be included in the experiment and the testing conditions is essential for the predictions made through the laboratory experiment to be valid under both manufacturing and field conditions.

For the polysilicon deposition case study, four batches of 50 wafers containing 3 test wafers were processed under both the optimum condition and under the starting conditions. The results are tabulated in Table 4.11. It is clear that the data agree very well with the predictions about the improvement in the S/N ratios and the deposition rate. So, we could adopt the optimum settings as the new process settings and proceed to implement these settings.

TABLE 4.10 PREDICTION USING THE ADDITIVE MODEL

	Starting Condition				Optimum Condition			
		Contribution† (dB)				Contribution† (dB)		
Factor	Setting	Surface Defects	Thickness	Deposition Rate	Setting	Surface Defects	Thickness	Deposition Rate
A*	A_2	−4.74	3.39	0.01	A_1	21.13	3.60	−5.36
B	B_2	−2.08	0.00	0.66	B_2	−2.08	0.00	0.66
C	C_1	6.33	2.87	−1.31	C_1	6.33	2.87	−1.31
D	D_3	−4.68	−3.35	1.49	D_3	−4.68	−3.35	1.49
E*	E_1	−6.16	0.00	0.00	E_2	4.82	0.00	0.00
F*	F_1	0.00	−4.48	0.00	F_2	0.00	2.15	0.00
Overall Mean		−45.36	31.52	34.12		−45.36	31.52	34.12
Total		−56.69	29.95	34.97		−19.84	36.79	29.60

* Indicates the factors whose levels are changed from the starting to the optimum conditions.
† By *contribution* we mean the deviation from the overall mean caused by the particular factor level.

TABLE 4.11 RESULTS OF VERIFICATION EXPERIMENT

		Starting Condition	Optimum Condition	Improvement
Surface Defects	rms	600/cm²	7/cm²	
	η	−55.6 dB	−16.9 dB	38.7 dB
Thickness	std. dev.*	0.028	0.013	
	η′	31.1 dB	37.7 dB	6.6 dB
Deposition Rate	rate	60 Å /min	35 Å /min	
	η″	35.6 dBam	30.9 dBam	−4.7 dBam

* Standard deviation of thickness is expressed as a fraction of the mean thickness.

Follow-up Experiments

Optimization of a process or a product need not be completed in a single matrix experiment. Several matrix experiments may have to be conducted in sequence before completing a product or process design. The information learned in one matrix experiment is used to plan the subsequent matrix experiments for achieving even more improvement in the process or the product. The factors studied in such subsequent experiments, or the levels of the factors, are typically different from those studied in the earlier experiments.

From the case-study data on the polysilicon deposition process, temperature stood out as the most important factor—both for quality and productivity. The experimental data showed that high temperature leads to excessive formation of surface defects and nonuniform thickness. This led to identifying the type of temperature controller as a potentially important control factor. The controller used first was an underdamped controller, and, consequently, during the initial period of deposition, the reactor temperature rose significantly above the steady-state set-point temperature. It was then decided to try a critically damped controller. Thus, an auxiliary experiment was conducted with two control factors: (1) the type of controller, and (2) the temperature setting. This experiment identified the critically damped controller as being significantly better than the underdamped one.

The new controller allowed the temperature setting to be increased to $T_0 - 10$ °C while keeping the surface defect count below 1 defect/unit area. The higher temperature also led to a deposition rate of 55Å/min rather than the 35Å/min that was observed in the initial verification experiment. Simultaneously, a standard deviation of thickness equal to 0.007 times the mean thickness was achieved.

Range of Applicability

In any development activity, it is highly desirable that the conclusions continue to be valid when we advance to a new generation of technology. In the case study of the polysilicon deposition process, this means that having developed the process with 4-inch wafers, we would want it to be valid when we advance to 5-inch wafers. The process developed for one application should be valid for other applications. Processes and products developed by the Robust Design method generally possess this characteristic of design transferability. In the case study, going from 4-inch wafers to 5-inch wafers was achieved by making minor changes dictated by the thermal capacity calculations. Thus, a significant amount of development effort was saved in transferring the process to the reactor that handled 5-inch wafers.

4.9 SUMMARY

Optimizing the product or process design means determining the best architecture, levels of control factors, and tolerances. Robust Design is a methodology for finding the optimum settings of control factors to make the product or process insensitive to noise factors. It involves eight major steps which can be grouped as planning a matrix experiment to determine the effects of the control factors (Step 1 through 5), conducting the matrix experiment (Step 6), and analyzing and verifying the results (Steps 7 and 8).

- **Step 1**. Identify the main function, side effects and failure modes. This step requires engineering knowledge of the product or process and the customer's environment.

- **Step 2**. Identify noise factors and testing conditions for evaluating the quality loss. The testing conditions are selected to capture the effect of the more important noise factors. It is important that the testing conditions permit a consistent estimation of the sensitivity to noise factors for any combination of control factor levels. In the polysilicon deposition case study, the effect of noise factors was captured by measuring the quality characteristics at three specific locations on each of three wafers, appropriately placed along the length of the tube. Noise orthogonal array and compound noise factor are two common techniques for constructing testing conditions. These techniques are discussed in Chapter 8.

- **Step 3**. Identify the quality characteristic to be observed and the objective function to be optimized. Guidelines for selecting the quality characteristic and the objective function, which is generically called S/N ratio, are given in Chapters 5 and 6. The common temptation of using the percentage of products that meet the specification as the objective function to be optimized should be avoided. It leads to orders of magnitude reduction in efficiency of experimentation. While optimizing manufacturing processes, an appropriate throughput characteristic should also be studied along with the quality characteristics because the economics of the process is determined by both of them.

- **Step 4**. Identify the control factors and their alternate levels. The more complex a product or a process, the more control factors it has and vice versa. Typically, six to eight control factors are chosen at a time for optimization. For each control factor two or three levels are selected, out of which one level is usually the starting level. The levels should be chosen sufficiently far apart to cover a wide experimental region because sensitivity to noise factors does not usually change with small changes in control factor settings. Also, by choosing a wide experimental region, we can identify good regions, as well as bad regions, for control factors. Chapter 6 gives additional guidelines for choosing control factors and their levels. In the polysilicon deposition case study, we investigated three levels each of six control factors. One of these factors (cleaning method) had discrete levels. For four of the factors the ratio of the largest to the smallest levels was between three and five.

- **Step 5**. Design the matrix experiment and define the data analysis procedure. Using orthogonal arrays is an efficient way to study the effect of several control factors simultaneously. The factor effects thus obtained are valid over the experimental region and it provides a way to test for the additivity of the factor effects. The experimental effort needed is much smaller when compared to other methods of experimentation, such as guess and test (trial and error), one factor at a time, and full factorial experiments. Also, the data analysis is easy when orthogonal arrays are used. The choice of an orthogonal array for a particular project depends on the number of factors and their levels, the convenience of changing the levels of a particular factor, and other practical considerations. Methods for constructing a suitable orthogonal array are given in Chapter 7. The orthogonal array L_{18}, consisting of 18 experiments, was used for the polysilicon deposition study. The array L_{18} happens to be the most commonly used array because it can be used to study up to seven 3-level and one 2-level factors.

- **Step 6**. Conduct the matrix experiment. Levels of several control factors must be changed when going from one experiment to the next in a matrix experiment. Meticulousness in correctly setting the levels of the various control factors is essential—that is, when a particular factor has to be at level 1, say, it should not be set at level 2 or 3. However, one should not worry about small perturbations that are inherent in the experimental equipment. Any erroneous experiments or missing experiments must be repeated to complete the matrix. Errors can be avoided by preparing the experimenter's log and data sheets prior to conducting the experiments. This also speeds up the conduct of the experiments significantly. The 18 experiments for the polysilicon deposition case study were completed in 9 days.

- **Step 7**. Analyze the data, determine optimum levels for the control factors, and predict performance under these levels. The various steps involved in analyzing the data resulting from matrix experiments are described in Chapter 3. S/N ratios and other summary statistics are first computed for each experiment. (In Robust Design, the primary focus is on maximizing the S/N ratio.) Then, the factor effects are computed and ANOVA performed. The factor effects, along with their confidence intervals, are plotted to assist in the selection of their optimum levels. When a product or a process has multiple quality characteristics, it may become necessary to make some trade-offs while choosing the optimum factor levels. The observed factor effects together with the quality loss function can be used to make rational trade-offs. In the polysilicon case study, the data analysis indicated that levels of three factors—deposition temperature (A), settling time (E), and cleaning method (F)—be changed, while the levels of the other five factors be kept at their starting levels.

- **Step 8**. Conduct the verification (confirmation) experiment and plan future actions. The purpose of this final and crucial step is to verify that the optimum conditions suggested by the matrix experiments do indeed give the projected improvement. If the observed and the projected improvements match, we adopt the suggested optimum conditions. If not, then we conclude that the additive model underlying the matrix experiment has failed, and we find ways to correct that problem. The corrective actions include finding better quality characteristics, or signal-to-noise ratios, or different control factors and levels, or studying a few specific interactions among the control factors. Evaluating the improvement in quality loss, defining a plan for implementing the results, and deciding whether another cycle of experiments is needed are also a part of this final step of Robust Design. It is quite common for a product or process design to require more than one cycle of Steps 1 through 8 for achieving needed quality and cost improvement. In the polysilicon deposition case study, the verification experiment confirmed the optimum conditions suggested by the data analysis. In a follow up Robust Design cycle, two control factors were studied—deposition temperature and type of temperature controller. The final optimum process gave nearly two orders of magnitude reduction in surface defects and a 4-fold reduction in the standard deviation of the thickness of the polysilicon layer.

Chapter 5

SIGNAL-TO-NOISE RATIOS

The concept of quadratic loss function introduced in Chapter 2 is ideally suited for evaluating the quality level of a product as it is shipped by a supplier to a customer. "As shipped" quality means that the customer would use the product without any adjustment to it or to the way it is used. Of course, the customer and the supplier could be two departments within the same company.

A few common variations of the quadratic loss function were given in Chapter 2. Can we use the quadratic loss function directly for finding the best levels of the control factors? What happens if we do so? What objective function should we use to minimize the sensitivity to noise? We examine these and other related questions in this chapter. In particular, we describe the concepts behind the signal-to-noise (S/N) ratio and the rationale for using it as the objective function for optimizing a product or process design. We identify a number of common types of engineering design problems and describe the appropriate S/N ratios for these problems. We also describe a procedure that could be used to derive S/N ratios for other types of problems. This chapter has six sections:

- Section 5.1 discusses the analysis of the polysilicon thickness uniformity. Through this discussion, we illustrate the disadvantages of direct minimization of the quadratic loss function and the benefits of using S/N ratio as the objective function for optimization.

- Section 5.2 presents a general procedure for deriving the S/N ratio.

- Section 5.3 describes common *static problems* (where the target value for the quality characteristic is fixed) and the corresponding S/N ratios.

- Section 5.4 discusses common *dynamic problems* (where the quality characteristic is expected to follow the signal factor) and the corresponding S/N ratios.

- Section 5.5 describes the accumulation analysis method for analyzing ordered categorical data.

- Section 5.6 summarizes the important points of this chapter.

5.1 OPTIMIZATION FOR POLYSILICON LAYER THICKNESS UNIFORMITY

One of the two quality characteristics optimized in the case study of the polysilicon deposition process in Chapter 4 was the thickness of the polysilicon layer. Recall that one of the goals was to achieve a uniform thickness of 3600 Å . More precisely, the experimenters were interested in minimizing the variance of thickness while keeping the mean on target. The objective of many robust design projects is to achieve a particular target value for the quality characteristic under all noise conditions. These types of projects were previously referred to as nominal-the-best type problems. The detailed analysis presented in this section will be helpful in formulating such projects. This section discusses the following issues:

- Comparison of the quality of two process conditions

- Relationship between S/N ratio and quality loss after adjustment (Q_a)

- Optimization for different target thickness

- Interaction induced by the wrong choice of objective function

- Identification of a scaling factor

- Minimization of standard deviation and mean separately

Comparing the Quality of Two Process Conditions

Suppose we are interested in determining which is a preferred temperature setting, T_0 °C or $(T_0 + 25)$ °C, for achieving uniform thickness of the polysilicon layer around the target thickness of 3600 Å . We may attempt to answer this question by running a number of batches under the two temperature settings while keeping the other control factors fixed at certain levels. Suppose the observed mean thickness and standard deviation of thickness for these two process conditions are as given in Table 5.1. Although no experiments were actually conducted under these conditions, the data in Table 5.1 are realistic based on experience with the process. This is also true for all other data used in this section. Note that under temperature T_0 °C, the mean thickness is 1800 Å , which is far away from the target, but the standard deviation is small. Whereas under temperature $(T_0 + 25)$ °C, the mean thickness is 3400 Å , which is close to the target, but the standard deviation is large. As we observe here, *it is very typical for both the mean and standard deviation to change when we change the level of a factor.*

TABLE 5.1 EFFECT OF TEMPERATURE ON THICKNESS UNIFORMITY

Expt. No.	Temperature (°C)	Mean Thickness (μ)* (Å)	Standard Deviation (σ) (Å)	Q† (Å)^2
1	T_0	1800	32	3.241×10^6
2	$T_0 + 25$	3400	200	8.000×10^4

* Target mean thickness = μ_0 = 3600 Å

† $Q = (\mu - \mu_0)^2 + \sigma^2$

From the data presented in Table 5.1, which temperature setting can we recommend? Since both the mean and standard deviation change when we change the temperature, we may decide to use the quadratic loss function to select the better temperature setting. For a given mean, μ, and standard deviation, σ, the quality loss without adjustment, denoted by Q, is given by

$$Q = \text{quality loss without adjustment} = k\left[(\mu - 3600)^2 + \sigma^2\right] \qquad (5.1)$$

where k is the quality loss coefficient. Note that throughout this chapter we ignore the constant k (that is, set it equal to 1) because it has no effect on the choice of optimum levels for the control factors. The quality loss under T_0 °C is 3.24×10^6, while under $(T_0 + 25)$ °C it is 8.0×10^4. Thus, we may conclude that $(T_0 + 25)$ °C is the better temperature setting. But, is that really a correct conclusion?

Recall that the deposition time is a scaling factor for the deposition process—that is, for any fixed settings of all other control factors, the polysilicon thickness at the various points within the reactor is proportional to the deposition time. Of course, the proportionality constant, which is the same as the deposition rate, could be different at different locations within the reactor. This is what leads to the variance, σ^2, of the polysilicon thickness. We can use this knowledge of the scaling factor to estimate the quality loss after adjusting the mean on target.

For T_0 °C temperature, we can attain the mean thickness of 3600 Å by increasing the deposition time by a factor of 3600/1800 = 2.0. Correspondingly, the standard deviation would also increase by the factor of 3600/1800 to 64 Å . Thus, the estimated quality loss after adjusting the mean is 4.1×10^3. Similarly, for $(T_0 + 25)$ °C we can obtain 3600 Å thickness by increasing the deposition time by a factor of 3600 / 3400, which would result in a standard deviation of 212 Å . Thus, the estimated quality loss after adjusting the mean is 4.49×10^4. From these calculations it is clear that when the mean is adjusted to be on target, the quality loss for T_0 °C is an order of magnitude smaller than the quality loss for $(T_0 + 25)$ °C; that is, the sensitivity to noise is much less when the deposition temperature is T_0 °C as opposed to $(T_0 + 25)$ °C. Hence, T_0 °C is the preferred temperature setting.

A decision based on quality loss without adjustment (Q) is influenced not only by the sensitivity to noise (σ^2), but also by the deviation from the target mean ($\mu - \mu_0$). Often, such a decision is heavily influenced, if not dominated, by the deviation from the target mean. As a result, we risk the possibility of not choosing the factor level that minimizes sensitivity to noise. This, of course, is clearly undesirable. But when we compute the quality loss after adjustment, denoted by Q_a, for all practical purposes we eliminate the effect of change in mean. In fact, it is a way of isolating the sensitivity to noise factors. Thus, a decision based on Q_a minimizes the sensitivity to noise, which is what we are most interested in during robust design.

Relationship between S/N Ratio and Q_a

The general formula for computing the quality loss after adjustment for the polysilicon thickness problem, which is a nominal-the-best type problem, can be derived as follows: If the observed mean thickness is μ, we have to increase the deposition time by a factor of μ_0/μ to get the mean thickness on target. The predicted standard deviation after adjusting the mean on target is $(\mu_0/\mu) \, \sigma$, where σ is the observed standard deviation. So, we have

$$Q_a = \text{quality loss after adjustment} = k \left[\frac{\mu_0}{\mu} \, \sigma \right]^2 . \qquad (5.2)$$

We can rewrite Equation (5.2) as follows:

$$Q_a = k\mu_0^2 \left[\frac{\sigma^2}{\mu^2} \right]. \qquad (5.3)$$

Since in a given project k and μ_0 are constants, we need to focus our attention only on (μ^2/σ^2). We call (μ^2/σ^2) the S/N ratio because σ^2 is the effect of noise factors and μ is the desirable part of the thickness data. Maximizing (μ^2/σ^2) is equivalent to minimizing the quality loss after adjustment, given by Equation (5.3), and also equivalent to minimizing sensitivity to noise factors.

For improved additivity of the control factor effects, it is common practice to take log transform of (μ^2/σ^2) and express the S/N ratio in decibels,

$$\eta = 10 \, \log_{10} \left[\frac{\mu^2}{\sigma^2} \right]. \tag{5.4}$$

Although it is customary to refer to both (μ^2/σ^2) and η as the S/N ratio, it is clear from the context which one we mean. The range of values of (μ^2/σ^2) is $(0, \infty)$, while the range of values of η is $(-\infty, \infty)$. Thus, in the log domain, we have better additivity of the effects of two or more control factors. Since log is a monotone function, maximizing (μ^2/σ^2) is equivalent to maximizing η.

Optimization for Different Target Thicknesses

Using the S/N ratio rather than the mean square deviation from target as an objective function has one additional advantage. Suppose for a different application of the polysilicon deposition process, such as manufacturing a new code of microchips, we want to have 3000 Å target thickness. Then, the optimum conditions obtained by maximizing the S/N ratio would still be valid, except for adjustment of the mean. However, the same cannot be said if we used the mean square deviation from target as the objective function. We would have to perform the optimization again.

The problem of minimizing the variance of thickness while keeping the mean on target is a problem of constrained optimization. As discussed in Appendix B, by using the S/N ratio, the problem can be converted into an unconstrained optimization problem that is much easier to solve. The property of unconstrained optimization is the basis for our ability to separate the actions of minimizing sensitivity to noise factors by maximizing the S/N ratio and the adjustment of mean thickness on target.

When we advance from one technology of integrated circuit manufacturing to a newer technology, we must produce thinner layers, print and etch smaller width lines, etc. With this in mind, it is crucial that we focus our efforts on reducing sensitivity to noise by optimizing the S/N ratio. The mean can then be adjusted to meet the desired target. This flexible approach to process optimization is needed not only for integrated circuit manufacturing, but also for virtually all manufacturing processes and optimization of all product designs.

During product development, the design of subsystems and components must proceed in parallel. Even though the target values for various characteristics of the subsystems and components are specified at the beginning of the development activity, it often becomes necessary to change the target values as more is learned about the product. Optimizing the S/N ratio gives us the flexibility to change the target later in the development effort. Also, the reusability of the subsystem design for other applications is greatly enhanced. Thus, by using the S/N ratio we improve the overall productivity of the development activity.

Interactions Induced by Wrong Choice of Objective Function

Using the quality loss without adjustment as the objective function to be optimized can also lead to unnecessary interactions among the control factors. To understand this point, let us consider again the data in Table 5.1. Suppose the deposition time for the two experiments in Table 5.1 was 36 minutes. Now suppose we conducted two more experiments with 80 minutes of deposition time and temperatures of T_0 °C and $(T_0 + 25)$ °C. Let the data for these two experiments be as given in Table 5.2. For ease of comparison, the data from Table 5.1 are also listed in Table 5.2.

TABLE 5.2 INTERACTIONS CAUSED BY THE MEAN

Expt. No.	Temperature (°C)	Deposition Time (min)	Mean Thickness (μ)* ($\mathring{A}$)	Standard Deviation (σ) ($\mathring{A}$)	Q† ($\mathring{A}$)^2	Q_a‡ ($\mathring{A}$)^2
1	T_0	36	1800	32	3.241×10^6	4.096×10^3
2	$T_0 + 25$	36	3400	200	8.000×10^4	4.484×10^4
3	T_0	80	4000	70	1.649×10^5	3.969×10^3
4	$T_0 + 25$	80	7550	440	15.796×10^6	4.402×10^4

* Target mean thickness = $\mu_0 = 3600 \mathring{A}$

† $Q = (\mu - \mu_0)^2 + \sigma^2$

‡ $Q_a = \mu_0^2 \left[\dfrac{\sigma^2}{\mu^2} \right]$

The quality loss without adjustment is plotted as a function of temperature for the two values of deposition time in Figure 5.1(a). We see that for 36 minutes of deposition time, the $(T_0 + 25)$ °C is the preferred temperature, whereas for 80 minutes of deposition time the preferred temperature is T_0 °C. Such opposite conclusions about the optimum levels of control factors (called *interactions*) are a major source of confusion and inefficiency in experimentation for product or process design improvement.

Not only is the estimation of interaction expensive but the estimation might not yield the true optimum settings for the control factors—that is, if there are strong antisynergistic interactions among the control factors, we risk the possibility of choosing a wrong combination of factor levels for the optimum conditions. In this example, based on Q, we would pick the combination of $(T_0 + 25)$ °C and 36 minutes as the best combination. But, if we use the S/N ratio or the Q_a as the objective function, we would unambiguously conclude that T_0 °C is the preferred temperature [see Figure 5.1(b)].

(a) When Q is the objective
function, the control factors,
temperature and time, have
strong antisynergistic
interaction.

(b) When Q_a is the objective
function, there is no interaction
between temperature and time.
Here, since time is a scaling factor,
the curves for 36 min. and
80 min. deposition time are
almost overlapping.

(c) From this figure we see that
much of the interaction in (a)
is caused by the deviation of
the mean from the target.

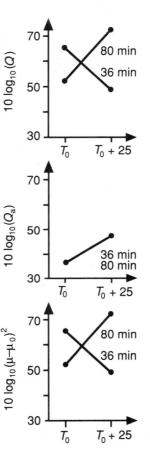

Figure 5.1   Interactions caused by the mean.

The squared deviation of the mean from the target thickness is a component of the objective function Q [see Equation (5.1)]. This component is plotted in Figure 5.1(c). From the figure it is obvious that the interaction revealed in Figure 5.1(a) is primarily caused by this component. The objective function Q_a does not have the squared deviation of the mean from the target as a component. Consequently, the corresponding interaction, which unnecessarily complicates the decision process, is eliminated.

In general, if we observe that for a particular objective function the interactions among the control factors are strong, we should look for the possibility that the objective function may have been selected incorrectly. The possibility exists that the objective function did not properly isolate the effect of noise factors and that it still has the deviation of the product's mean function from the target as a component.

Identification of a Scaling Factor

In the polysilicon deposition case study, the deposition time is an easily identified scaling factor. However, in many situations where we want to obtain mean on target, the scaling factor cannot be identified readily. How should we determine the best settings of the control factors in such situations?

It might, then, be tempting to use the mean squared deviation from the target as the objective function to be minimized. However, as explained earlier, minimizing the mean squared deviation from the target can lead to wrong conclusions about the optimum levels for the control factors; so, the temptation should be avoided. Instead, we should begin with an assumption that a scaling factor exists and identify such a factor through experiments.

The objective function to be maximized, namely η, can be computed from the observed μ and σ without knowing which factor is a scaling factor. Also, the scaling operation does not change the value of η. Thus, the process of discovering a scaling factor and the optimum levels for the various control factors is a simple one. It consists of determining the effect of every control factor on η and μ, and then classifying these factors as follows:

1. *Factors that have a significant effect on η.* For these factors, we should pick the levels that maximize η.

2. *Factors that have a significant effect on μ but practically no effect on η.* Any one of these factors can serve as a scaling factor. We use one such factor to adjust the mean on target. We are generally successful in finding at least one scaling factor. However, sometimes we must settle for a factor that has a small effect on η as a scaling factor.

3. *Factors that have no effect on η and no effect on μ.* These are neutral factors and we can choose their best levels from other considerations such as ease of operation or cost.

Minimizing Standard Deviation and Mean Separately

Another way to approach the problem of minimizing variance with the constraint that the mean should be on target is, first, to minimize standard deviation while ignoring the mean, and, then, bring the mean on target without affecting the standard deviation by changing a suitable factor. The difficulty with this approach is that often we *cannot find a factor that can change the mean over a wide range without affecting the standard deviation.* This can be understood as follows: In these problems, when the mean is zero, the standard deviation is also zero. However, for all other mean values, the standard deviation cannot be identically zero. Thus, whenever a factor changes the mean, it also affects the standard deviation. Also, an attempt to minimize standard deviation without paying attention to the mean drives both the standard deviation and the mean to zero, which is not a worthwhile solution. Therefore, we should not try to minimize σ without paying attention to the mean. However, we can almost always find a scaling factor. Thus, an approach where we maximize the S/N ratio leads to useful solutions.

Note that the above discussion pertains to the class of problems called nominal-the-best type problems, of which polysilicon thickness uniformity is an example. A class of problems called signed-target type problems where it is appropriate to first minimize variance and then bring the mean on target is described in Section 5.3.

5.2 EVALUATION OF SENSITIVITY TO NOISE

Let us now examine the general problem of evaluating sensitivity to noise for a dynamic system. Recall that in a dynamic system the quality characteristic is expected to follow the signal factor. The ideal function for many products can be written as

$$y = M \tag{5.5}$$

where y is the quality characteristic (or the observed response) and M is the signal (or the command input). In this section we discuss the evaluation of sensitivity to noise for such dynamic systems. For specificity, suppose we are optimizing a servomotor (a device such as an electric motor whose movement is controlled by a signal from a command device) and that y is the displacement of the object that is being moved by the servomotor and M specifies the desired displacement. To determine the sensitivity of the servomotor, suppose we use the signal values $M_1, M_2, \cdots, M_m$; and for each signal value, we use the noise conditions $x_1, x_2, \cdots, x_n$. Let y_{ij} denote the observed displacement for a particular value of control factor settings, $\mathbf{z} = (z_1, z_2, \cdots, z_q)^T$, when the signal is M_i and noise is x_j. Representative values of y_{ij} and the ideal function are shown in Figure 5.2. The average quality loss, $Q(\mathbf{z})$, associated with the control factor settings, $\mathbf{z}$, is given by

$$Q(\mathbf{z}) = \frac{k}{mn} \sum_{i=1}^{m} \sum_{j=1}^{n} (y_{ij} - M_i)^2. \tag{5.6}$$

As shown by Figure 5.2, $Q(\mathbf{z})$ includes not only the effect of noise factors but also the deviation of the mean function from the ideal function. In practice, $Q(\mathbf{z})$ could be dominated by the deviation of the mean function from the ideal function. Thus, the direct minimization of $Q(\mathbf{z})$ could fail to achieve truly minimum sensitivity to noise. It could lead simply to bringing the mean function on target, which is not a difficult problem in most situations anyway. Therefore, *whenever adjustment is possible, we should minimize the quality loss after adjustment.*

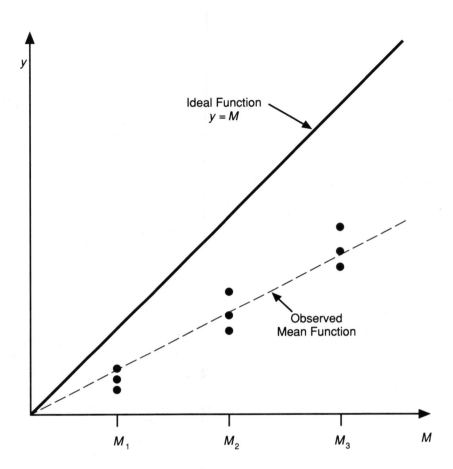

Figure 5.2   Evaluation of sensitivity to noise.

For the servomotor, it is possible to adjust a gear ratio so that, referring to Figure 5.2, the slope of the observed mean function can be made equal to the slope of the ideal function. Let the slope of the observed mean function be β. By changing the gear ratio we can change every displacement y_{ij} to $v_{ij} = (1/\beta)y_{ij}$. This brings the mean function on target.

For the servomotor, change of gear ratio leads to a simple linear transformation of the displacement y_{ij}. In some products, however, the adjustment could lead to a more complicated function between the adjusted value v_{ij} and the unadjusted value y_{ij}. For a general case, let the effect of the adjustment be to change each y_{ij} to a value $v_{ij} = h_R (y_{ij})$, where the function h_R defines the adjustment that is indexed by a parameter R. After adjustment, we must have the mean function on target—that is, the errors

$(v_{ij} - M_i)$ must be orthogonal to the signal M_i. Mathematically, the requirement of orthogonality can be written as

$$\sum_{i=1}^{m} \sum_{j=1}^{n} (v_{ij} - M_i) M_i = 0. \tag{5.7}$$

Equation (5.7) can be solved to determine the best value of R for achieving the mean function on target. Then the quality loss after adjustment, $Q_a(\mathbf{z})$, can be evaluated as follows:

$$Q_a(\mathbf{z}) = \frac{k}{mn} \sum_{i=1}^{m} \sum_{j=1}^{n} (v_{ij} - M_i)^2. \tag{5.8}$$

The quantity $Q_a(\mathbf{z})$ is a measure of sensitivity to noise. It does not contain any part that can be reduced by the chosen adjustment process. However, any systematic part of the relationship between y and M that cannot be adjusted is included in $Q_a(\mathbf{z})$. [For the servomotor, the nonlinearity (2nd, 3rd, and higher order terms) of the relationship between y and M are contained in $Q_a(\mathbf{z})$.] Minimization of $Q_a(\mathbf{z})$ makes the design robust against the noise factors and reduces the nonadjustable part of the relationship between y and M. Any control factor that has an effect on y_{ij} but has no effect on $Q_a(\mathbf{z})$ can be used to adjust the mean function on target without altering the sensitivity to noise, which has already been minimized. Such a control factor is called an *adjustment factor*.

It is easy to verify that minimization of $Q_a(\mathbf{z})$, followed by adjusting the mean function on target using an adjustment factor, is equivalent to minimizing $Q(\mathbf{z})$ subject to the constraint that the mean function is on target. This optimization procedure is called a *two-step procedure* for obvious reasons. For further discussion of the 2-step procedure and the S/N ratios, see Taguchi and Phadke [T6], Phadke and Dehnad [P4], Leon, Shoemaker, and Kackar [L2], Nair and Pregibon [N2], and Box [B1].

It is important to be able to predict the combined effect of several control factors from the knowledge of the effects of the individual control factors. The natural scale of $Q_a(\mathbf{z})$ is not suitable for this purpose because it could easily give us a negative prediction for $Q_a(\mathbf{z})$ which is absurd. By using the familiar decibel scale, we not only avoid the negative prediction, but also improve the additivity of the factor effects. Thus, to minimize the sensitivity to noise factors, we maximize η which is given by

$$\eta = -10 \log_{10} Q_a(\mathbf{z}). \tag{5.9}$$

Note that the constant k in $Q_a(\mathbf{z})$ and sometimes some other constants are generally ignored because they have no effect on the optimization.

Following Taguchi, we refer to η as the S/N ratio. In the polysilicon deposition example discussed in Section 5.1, we saw that $Q_a \propto (\mu^2/\sigma^2)$, where σ^2 is the effect of the noise factors, and μ is the desirable part of the thickness data. Thus Q_a is the ratio of the power of the signal (the desirable part) to the power of the noise (the undesirable part). As will be seen through the cases discussed in the subsequent sections of this chapter, whenever a scaling type of an adjustment factor exists, Q_a takes the form of a ratio of the power of the signal (the desirable part of the response) to the power of the noise (the undesirable part of the response). Therefore, Q_a and η are both referred to as the S/N ratio. As a matter of convention, we call Q_a and η the S/N ratio, even in other cases where "ratio" form is not that apparent.

The general optimization strategy can be summarized as follows:

1. Evaluate the effects of the control factors under consideration on η and on the mean function.

2. For the factors that have a significant effect on η, select levels that maximize η.

3. Select any factor that has no effect on η but a significant effect on the mean function as an adjustment factor. In practice, we must sometimes settle for a factor that has a small effect on η but a significant effect on the mean function as an adjustment factor. Use the adjustment factor to bring the mean function on target. Adjusting the mean function on target is the main quality control activity in manufacturing. It is needed because of changing raw material, varying processing conditions, etc. Thus, finding an adjustment factor that can be changed conveniently during manufacturing is important. However, finding the level of the adjustment factor that brings the mean precisely on target during product or process design is not important.

4. For factors that have no effect on η and the mean function, we can choose any level that is most convenient from the point of view of other considerations, such as other quality characteristics and cost.

What adjustment is meaningful in a particular engineering problem and what factor can be used to achieve the adjustment depend on the nature of the particular problem. Subsequent sections discuss several common engineering problems and derive the appropriate S/N ratios using the results of this section.

5.3 S/N RATIOS FOR STATIC PROBLEMS

Finding a correct objective function to maximize in an engineering design problem is very important. Failure to do so, as we saw earlier, can lead to great inefficiencies in experimentation and even wrong conclusions about the optimum levels. The task of finding what adjustments are meaningful in a particular problem and determining the correct S/N ratio is not always easy. Here, we describe some common types of static problems and the corresponding S/N ratios.

Minimizing the surface defect count and achieving target thickness in polysilicon deposition are both examples of static problems. In each case, we are interested in a fixed target, so that the signal factor is trivial, and for all practical purposes, we can say it is absent. In contrast, the design of an electrical amplifier is a dynamic problem in which the input signal is the signal factor and our requirement is to make the output signal proportional to the input signal. The tracking of the input signal by the output signal makes it a dynamic problem. We discuss dynamic problems in Section 5.4.

Static problems can be further characterized by the nature of the quality characteristic. Recall that the response we observe for improving quality is called quality characteristic. The classification of static problems is based on whether the quality characteristic is:

- Continuous or discrete

- Scalar, vector, or curve (such as frequency response function)

- Positive or covers the entire real line

- Such that the target value is extreme or finite

Commonly encountered types of static problems and the corresponding S/N ratios are described below (see also Taguchi and Phadke [T6] and Phadke and Dehnad [P4]). In these problems, the signal factor takes only one value. Thus, we denote by $y_1, y_2, \cdots, y_n$ the n observations of the quality characteristic under different noise conditions.

(a) Smaller-the-better Type Problem

Here, the quality characteristic is continuous and nonnegative—that is, it can take any value from 0 to ∞. Its most desired value is zero. Such problems are characterized by the absence of a scaling factor or any other adjustment factor. The surface defect count is an example of this type of problem. Note that for all practical purposes we can treat this count as a continuous variable.

Another example of a smaller-the-better type problem is the pollution from a power plant. One might say that we can reduce the total pollutants emitted by reducing the power output of the plant. So why not consider the power output as an adjustment factor? However, reducing pollution by reducing power consumption does not signify any quality improvement for the power plant. Hence, it is inappropriate to think of the power output as an adjustment factor. In fact, we should consider the pollution per megawatt-hour of power output as the quality characteristic to be improved instead of the pollution itself.

Additional examples of smaller-the-better type problems are electromagnetic radiation from telecommunications equipment, leakage current in integrated circuits, and corrosion of metals and other materials.

Because there is no adjustment factor in these problems, we should simply minimize the quality loss without adjustment—that is, we should minimize

$$Q = k(\text{mean square quality characteristic})$$

$$= k \left[\frac{1}{n} \sum_{i=1}^{n} y_i^2 \right]. \tag{5.10}$$

Minimizing Q is equivalent to maximizing η defined by the following equation,

$$\eta = -10 \log_{10} (\text{mean square quality characteristic})$$

$$= -10 \log_{10} \left[\frac{1}{n} \sum_{i=1}^{n} y_i^2 \right]. \tag{5.11}$$

Note that we have ignored the constant k and expressed the quality loss in the decibel scale.

In this case the signal is constant, namely to make the quality characteristic equal to zero. Therefore, the S/N ratio, η, measures merely the effect of noise.

(b) Nominal-the-best Type Problem

Here, as in smaller-the-better type problem, the quality characteristic is continuous and nonnegative—that is, it can take any value from 0 to ∞. Its target value is nonzero and finite. For these problems when the mean becomes zero, the variance also becomes zero. Also, for these problems we can find a scaling factor that can serve as an adjustment factor to move the mean on target.

This type of problem occurs frequently in engineering design. We have already discussed the problem in great detail with particular reference to achieving target thickness in polysilicon deposition. The objective function to be maximized for such problems is

$$\eta = 10 \log_{10} \frac{\mu^2}{\sigma^2} \tag{5.12}$$

where

$$\mu = \frac{1}{n} \sum_{i=1}^{n} y_i \ \text{ and } \ \sigma^2 = \frac{1}{n-1} \sum_{i=1}^{n} (y_i - \mu)^2. \qquad (5.13)$$

In some situations, the scaling factor can be identified readily through engineering expertise. In other situations, we can identify a suitable scaling factor through experimentation.

The optimization of the nominal-the-best problems can be accomplished in two steps:

1. Maximize η or minimize sensitivity to noise. During this step we select the levels of control factors to maximize η while we ignore the mean.

2. Adjust the mean on target. During this step we use the adjustment factor to bring the mean on target without changing η.

Note that as we explained in Section 5.1, we should not attempt to minimize σ and then bring the mean on target.

(c) Larger-the-better Type Problem

Here, the quality characteristic is continuous and nonnegative, and we would like it to be as large as possible. Also, we do not have any adjustment factor. Examples of such problems are the mechanical strength of a wire per unit cross-section area, the miles driven per gallon of fuel for an automobile carrying a certain amount of load, etc. This problem can be transformed into a smaller-the-better type problem by considering the reciprocal of the quality characteristic. The objective function to be maximized in this case is given by

$$\eta = -10 \log_{10} \text{ (mean square reciprocal quality characteristic)}$$

$$= -10 \log_{10} \left[\frac{1}{n} \sum_{i=1}^{n} \frac{1}{y_i^2} \right]. \qquad (5.14)$$

The following questions are often asked about the larger-the-better type problems: Why do we take the reciprocal of a larger-the-better type characteristic and then treat it as a smaller-the-better type characteristic? Why do we not maximize the mean square quality characteristic? This can be understood from the following result from mathematical statistics:

$$\text{Mean square reciprocal quality characteristic} \cong \frac{1}{\mu^2} \left[1 + 3\frac{\sigma^2}{\mu^2} \right]$$

where μ and σ are the mean and variance of the quality characteristic. [Note that if y denotes the quality characteristic, then the mean square reciprocal quality characteristic is the same as the expected value of $(1/y)^2$.] Minimizing the mean square reciprocal quality characteristic implies maximizing μ and minimizing σ^2, which is the desired thing to do. However, if we were to try to maximize the mean square quality characteristic, which is equal to $(\mu^2 + \sigma^2)$, we would end up maximizing both μ and σ^2, which is not a desirable thing to do.

(d) Signed-target Type Problem

In this class of problems, the quality characteristic can take positive as well as negative values. Often, the target value for the quality characteristic is zero. If not, the target value can be made zero by appropriately selecting the reference value for the quality characteristic. Here, we can find an adjustment factor that can move the mean without changing the standard deviation. Note that signed-target problems are inherently different from smaller-the-better type problems, even though in both cases the best value is zero. In the signed-target problems, the quality characteristic can take positive as well as negative values whereas in the smaller-the-better type problems the quality characteristic cannot take negative values. The range of possible values for the quality characteristic also distinguishes signed-target problems from nominal-the-best type problems. There is one more distinguishing feature. In signed-target type problems when the mean is zero, the standard deviation is not zero, but in nominal-the-best type problems when the mean is zero, the standard deviation is also zero.

An example of signed-target problems is the dc offset voltage of a differential operational amplifier. The offset voltage could be positive or negative. If the offset voltage is consistently off zero, then we can easily compensate for it in the circuit that receives the output of the differential operational amplifier without affecting the standard deviation. The design of a differential operational amplifier is discussed in Chapter 8.

In such problems, the objective function to be maximized is given by

$$\eta = -10 \log_{10} \sigma^2$$

$$= -10 \log_{10} \left[\frac{1}{n-1} \sum_{i=1}^{n} (y_i - \mu)^2 \right]. \tag{5.15}$$

Note that this type of problem occurs relatively less frequently compared to the nominal-the-best type problems.

(e) Fraction Defective

This is the case when the quality characteristic, denoted by p, is a fraction taking values between 0 and 1. Obviously, the best value for p is zero. Also, there is no adjustment factor for these problems. When the fraction defective is p, on an average, we have to manufacture $1/(1-p)$ pieces to produce one good piece. Thus, for every good piece produced, there is a waste and, hence, a loss that is equivalent to the cost of processing $\{1/(1-p)-1\} = p/(1-p)$ pieces. Thus, the quality loss is given by Q,

$$Q = k \, \frac{p}{1-p} \tag{5.16}$$

where k is the cost of processing one piece. Ignoring k, we obtain the objective function to be maximized in the decibel scale as

$$\eta = -10 \log_{10} \left[\frac{p}{1-p} \right]. \tag{5.17}$$

Note that the range of possible values of Q is 0 to ∞, but the range of possible values of η is $-\infty$ to ∞. Therefore, the additivity of factor effects is better for η than for Q. The S/N ratio for the fraction-defective problems is the same as the familiar logit transform, which is commonly used in biostatistics for studying drug response.

(f) Ordered Categorical

Here, the quality characteristic takes ordered categorical values. For example, after a drug treatment we may observe a patient's condition as belonging to one of the following categories: worse, no change, good, or excellent. In this situation, the extreme category, excellent, is the most desired category. However, in some other cases, an intermediate category is the most desired category. For analyzing data from ordered categorical problems, we form cumulative categories and treat each category (or its compliment, as the case may be) as a fraction-defective type problem. We give an example of analysis of ordered categorical data in Section 5.5.

(g) Curve or Vector Response

As the name suggests, in this type of problem the quality characteristic is a curve or a vector rather than a single point. The treatment of this type of problem is described in Chapter 6 in conjunction with the design of an electrical filter and paper transport in copying machines. The basic strategy in these problems is to break them into several scalar problems where each problem is of one of the previously discussed types.

5.4 S/N RATIOS FOR DYNAMIC PROBLEMS

Dynamic problems have even more variety than static problems because of the many types of potential adjustments. Nonetheless, we use the general procedure described in Sections 5.1 and 5.3 to derive the appropriate objective functions or the S/N ratio. Dynamic problems can be classified according to the nature of the quality characteristic and the signal factor, and, also, the ideal relationship between the signal factor and the quality characteristic. Some common types of dynamic problems and the corresponding S/N ratios are given below (see also Taguchi [T1], Taguchi and Phadke [T6], and Phadke and Dehnad [P4]).

(a) Continuous-continuous (C-C)

Here, both the signal factor and the quality characteristic take positive or negative continuous values. When the signal is zero, that is, $M = 0$, the quality characteristic is also zero, that is, $y = 0$. The ideal function for these problems is $y = M$, and a scaling factor exists that can be used to adjust the slope of the relationship between y and M.

This is one of the most common types of dynamic problems. The servomotor example described in Section 5.2 is an example of this type. Some other examples are analog telecommunication, design of test sets (such as voltmeter and flow meter), and design of sensors (such as the crankshaft position sensor in an automobile).

We now derive the S/N ratio for the C-C type problems. As described in Section 5.2, let y_{ij} be the observed quality characteristic for the signal value M_i and noise condition x_j. The quality loss without adjustment is given by

$$Q = \frac{1}{mn} \sum_{i=1}^{m} \sum_{j=1}^{n} \left[y_{ij} - M_i \right]^2 .$$

The quality loss has two components. One is due to the deviation from linearity and the other is due to the slope being other than one. Of the two components, the latter can be eliminated by adjusting the slope. In order to find the correct adjustment for given control factor settings, we must first estimate the slope of the best linear relationship between y_{ij} and M_i. Consider the regression of y_{ij} on M_i given by

$$y_{ij} = \beta M_i + e_{ij} \tag{5.18}$$

where β is the slope and e_{ij} is the error. The slope β can be estimated by the least squares criterion as follows:

$$\frac{d}{d\beta}\left[\sum_{i=1}^{m}\sum_{j=1}^{n}(y_{ij} - \beta M_i)^2\right] = 0 \qquad (5.19)$$

that is

$$\sum_{i=1}^{m}\sum_{j=1}^{n}(y_{ij} - \beta M_i)M_i = 0 \qquad (5.20)$$

that is

$$\beta = \frac{\displaystyle\sum_{i=1}^{m}\sum_{j=1}^{n}(y_{ij}M_i)}{\displaystyle\sum_{i=1}^{m}\sum_{j=1}^{n}(M_i^2)}. \qquad (5.21)$$

Note that Equation (5.20) is nothing but a special case of the general Equation (5.7) for determining the best adjustment. Here, $h_R(y_{ij}) = (1/\beta)\,y_{ij} = v_{ij}$ and β is the same as the index R. Also note that the least squares criterion is analogous to the criterion of making the error $[(1/\beta)\,y_{ij} - M_i]$ orthogonal to the signal, M_i.

The quality loss after adjustment is given by

$$Q_a = \frac{k}{mn}\sum_{i=1}^{m}\sum_{j=1}^{n}(v_{ij} - M_i)^2$$

$$= \frac{k}{mn}\sum_{i=1}^{m}\sum_{j=1}^{n}\left[\frac{y_{ij}}{\beta} - M_i\right]^2$$

$$= \frac{k}{\beta^2}\cdot\frac{1}{mn}\sum_{i=1}^{m}\sum_{j=1}^{n}(y_{ij} - \beta M_i)^2$$

$$= k\cdot\frac{(mn-1)}{mn}\cdot\frac{\sigma_e^2}{\beta^2}$$

$$\cong k\cdot\frac{\sigma_e^2}{\beta^2} \qquad (5.21)$$

where the error variance, σ_e^2, is given by

$$\sigma_e^2 = \frac{1}{(mn-1)}\sum_{i=1}^{m}\sum_{j=1}^{n}(y_{ij} - \beta M_i)^2.$$

Minimizing Q_a is equivalent to maximizing η given by

$$\eta = 10 \log_{10} \frac{\beta^2}{\sigma_e^2}. \tag{5.22}$$

Note that β is the change in y produced by a unit change in M. Thus, β^2 quantifies the effect of signal. The denominator σ_e^2 is the effect of noise. Hence, η is called the S/N ratio. Note that σ_e^2 includes sensitivity to noise factors as well as the nonlinearity of the relationship between y and M. Thus, maximization of η leads to reduction in non-linearity along with the reduction in sensitivity to noise factors.

In summary, the C-C type problems are optimized by maximizing η given by Equation (5.22). After maximization of η, the slope is adjusted by a suitable scaling factor. Note that any control factor that has no effect on η but an appreciable effect on β can serve as a scaling factor.

Although we have shown the optimization for the target function $y = M$, it is still valid for all target functions that can be obtained by adjusting the slope—that is, the optimization is valid for any target functions of the form $y = \beta_0 M$ where β_0 is the desired slope.

Another variation of the C-C type target function is

$$y = \alpha_0 + \beta_0 M. \tag{5.23}$$

In this case, we must consider two adjustments: one for the intercept and the other for the slope. One might think of this as a *vector adjustment factor*. The S/N ratio to be maximized for this problem can be shown to be η, given by Equation (5.22). The two adjustment factors should be able to change the intercept and the slope, and should have no effect on η.

(b) Continuous-digital (C-D)

A temperature controller where the input temperature setting is continuous, while the output (which is the ON or OFF state of the heating unit) is discrete is an example of the C-D type problem. Such problems can be divided into two separate problems: one for the ON function and the other for the OFF function. Each of these problems can be viewed as a separate continuous-continuous or nominal-the-best type problem. The design of a temperature control circuit is discussed in detail in Chapter 9.

(c) Digital-continuous (D-C)

The familiar digital-to-analog converter is an example of the D-C case problem. Here again, we separate the problems of converting the 0 and 1 bits into the respective continuous values. The conversion of 0, as well as the conversion of 1, can be viewed as a nominal-the-best type static problem.

(d) Digital-digital (D-D)

Digital communication systems, computer operations, etc., where both the signal factor and the quality characteristic are digital, are examples of the D-D type problem. Here, the ideal function is that whenever 0 is transmitted, it should be received as 0, and whenever 1 is transmitted, it should be received as 1. Let us now derive an appropriate objective function for minimizing sensitivity to noise.

Here, the signal values for testing are $M_0 = 0$ and $M_1 = 1$. Suppose under certain settings of control factors and noise conditions, the probability of receiving 1, when 0 is transmitted, is p (see Table 5.3). Thus, the average value of the received signal, which is the same as the quality characteristic, is p and the corresponding variance is $p(1-p)$. Similarly, suppose the probability of receiving 0, when 1 is transmitted, is q. Then, the average value of the corresponding received signal is $(1-q)$ and the corresponding variance is $q(1-q)$. The ideal transmit-receive relationship and the observed transmit-receive relationship are shown graphically in Figure 5.3. Although the signal factor and the quality characteristic take only 0-1 values, for convenience we represent the transmit-receive relationship as a straight line. Let us now examine the possible adjustments.

TABLE 5.3 TRANSMIT-RECEIVE RELATIONSHIP FOR DIGITAL COMMUNICATION

		Probabilities Associated with the Received Signal		Properties of the Received Signal	
		0	**1**	**Mean**	**Variance**
Transmitted Signal	**0**	$1-p$	p	p	$p(1-p)$
	1	q	$1-q$	$1-q$	$q(1-q)$

It is well-known that a communication system is inefficient if the errors of transmitting 0 and 1 are unequal. More efficient transmission is achieved by making $p = q$. This can be accomplished by a leveling operation, an operation such as changing the threshold. The leveling operation can be conceptualized as follows: Underneath the transmission of a digital signal, there is a continuous signal such as voltage, frequency, or phase. If it is at all possible and convenient to observe the underlying continuous variable, we should prefer it. In that case, the problem can be classified as a C-D type and dealt with by the procedure described earlier. Here, we consider the situation when it is not possible to measure the continuous variable.

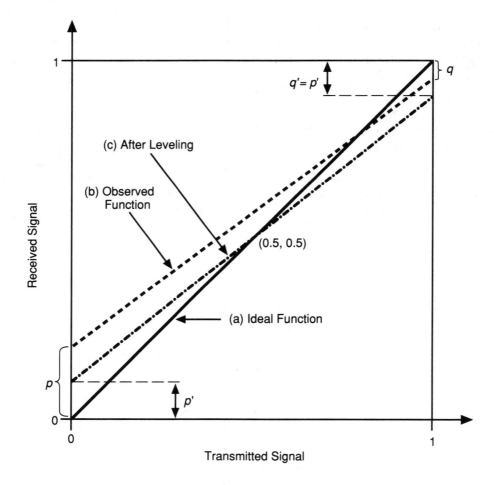

Figure 5.3   Digital communication.

Figure 5.4 shows possible distributions of the continuous variable received at the output terminal when 0 or 1 are transmitted. If the threshold value is R_1, the errors of 0 would be far more likely than the errors of 1. However, if the threshold is moved to

R_2, we would get approximately equal errors of 0 and 1. The effect of this adjustment is also to reduce the total error probability $(p + q)$.

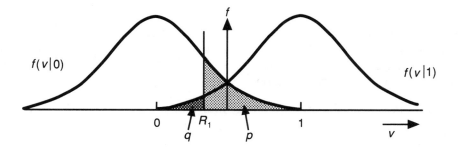

(a) When the threshold is at R_1, the error probabilities p and q are not equal.

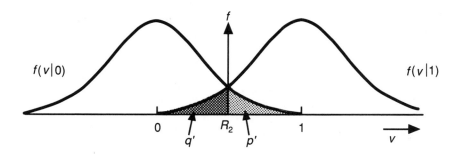

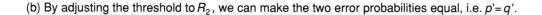

(b) By adjusting the threshold to R_2, we can make the two error probabilities equal, i.e. $p'=q'$.

Figure 5.4   Effect of leveling on error probabilities.

How does one determine p' (which is equal to q') corresponding to the observed error rates p and q? The relationship between p', p, and q will obviously depend on the continuous distribution. However, we are considering a situation where we do not have the ability to observe the distributions. Taguchi has suggested the use of the following relationship for estimating p' after equalization or leveling:

$$-2 \times 10 \log_{10} \left[\frac{p'}{1-p'} \right] = -10 \log_{10} \left[\frac{p}{1-p} \right] - 10 \log_{10} \left[\frac{q}{1-q} \right]. \qquad (5.24)$$

The two terms on the right hand side of Equation (5.24) are fraction-defective type S/N ratios for the separate problems of the errors of 0 and errors of 1. Equation (5.24) asserts that the effect of equalization is to make the two S/N ratios equal to the average of the S/N ratios before equalization.

We can rewrite Equation (5.24) as follows:

$$p' = \left[1 + \sqrt{\frac{1-p}{p} \cdot \frac{1-q}{q}} \right]^{-1}. \qquad (5.25)$$

Equation (5.25) provides an explicit expression computing p' from p and q.

The transmit-receive relationship after leveling is depicted by line c in Figure 5.3. When 0 is transmitted, the mean value of the received signal y_0', is p'. When 1 is transmitted, the mean value of the received signal, y_1', is $(1-q') = (1-p')$. In both cases the variance is $p'(1-p')$.

Both the lines a and c pass through the point (0.5, 0.5). But their slopes are not equal. The slope of line c is

$$\beta = \frac{0.5-p'}{0.5} = (1-2p') .$$

Thus, the quality loss after adjusting the slope is given by

$$Q_a = k \frac{\sigma_e^2}{\beta^2} = k \frac{p'(1-p')}{(1-2p')^2} . \qquad (5.26)$$

Thus, ignoring the constant k, the S/N ratio to be optimized is given by

$$\eta = 10 \log_{10} \frac{(1-2p')^2}{p'(1-p')} . \qquad (5.27)$$

Observe that $(1-2p')$ is the difference of the averages of the received signal when 0 and 1 are transmitted. The quantity $p'(1 - p')$ is the variance of the received signal. So η measures the ability of the communication system to discriminate between 0 and 1 at the receiving terminal.

The strategy to optimize a D-D system is to maximize η, and then use a control factor which has no effect on η, but can alter the ratio $p:q$ to equalize the two error probabilities.

Chemical separation processes can also be viewed as D-D type dynamic problems. For example, when making iron in a blast furnace, the goal is to separate the iron molecules from the impurities. Molten iron stays at the bottom and the impurities float as slag on top of the molten metal. Suppose $100\,p$ percent of iron molecules go into the slag and $100\,q$ percent of a particular impurity go in the molten metal. Minimization of p and q can be accomplished by maximizing the S/N ratio given by Equation (5.27). In this problem, however, the losses due to iron molecules going into the slag and impurities going into the iron are not equal. We may wish to make the $p:q$ ratio different from 1:1. The desired $p:q$ ratio can be accomplished during adjustment without altering η.

5.5 ANALYSIS OF ORDERED CATEGORICAL DATA

Data observed in many matrix experiments is ordered categorical because of the inherent nature of the quality characteristic or because of the convenience of the measurement technique. Let us consider the surface defect data from the polysilicon deposition case study of Chapter 4, which we will use to illustrate the analysis of ordered categorical data. Suppose, because of the inconvenience of counting the number of surface defects, the experimenters had decided to record the data in the following subjective categories, listed in progressively undersirable order: practically no surface defects, very few defects, some defects, many defects, and too many defects. For our illustration, we will take the observations listed in Table 4.4 (a) and associate with them categories I through V as follows:

$$
\begin{array}{rcl}
\text{I} & : & 0 - 3 \text{ defects} \\
\text{II} & : & 4 - 30 \text{ defects} \\
\text{III} & : & 31 - 300 \text{ defects} \\
\text{IV} & : & 301 - 1000 \text{ defects} \\
\text{V} & : & 1001 \text{ and more defects}
\end{array}
$$

Thus, among the nine observations of experiment 2, five belong to category I, two to category II, three to category III, and none to categories IV and V. The categorical data for the 18 experiments are listed in Table 5.4.

We will now describe Taguchi's accumulation analysis method [T7, T1], which is an effective method for determining optimum control factor settings in the case of ordered categorical data. (See Nair [N1] for an alternate method of analyzing ordered categorical data.) The first step is to define cumulative categories as follows:

$$
\begin{array}{lll}
(I) = I & : & 0 - 3 \text{ defects} \\
(II) = I+II & : & 0 - 30 \text{ defects} \\
(III) = I+II+III & : & 0 - 300 \text{ defects} \\
(IV) = I+II+III+IV & : & 0 - 1000 \text{ defects} \\
(V) = I+II+III+IV+V & : & 0 - \infty \text{ defects}
\end{array}
$$

The number of observations in the cumulative categories for the eighteen experiments are listed in Table 5.4. For example, the number of observations in the five cumulative categories for experiment 2 are 5, 7, 9, 9, and 9, respectively.

The second step is to determine the effects of the factor levels on the probability distribution by the defect categories. This is accomplished analogous to the determination of the factor effects described in Chapter 3. To determine the effect of temperature of level A_1, we identify the six experiments conducted at that level and sum the observations in each cumulative category as follows:

Cumulative Categories

	(I)	(II)	(III)	(IV)	(V)
Experiment 1	9	9	9	9	9
Experiment 2	5	7	9	9	9
Experiment 3	1	1	7	9	9
Experiment 10	9	9	9	9	9
Experiment 11	8	9	9	9	9
Experiment 12	2	5	8	8	9
Total	34	40	51	53	54

The number of observations in the five cumulative categories for every factor level are listed in Table 5.5. Note that the entry for the cumulative category (V) is equal to the total number of observations for the particular factor level and that entry is uniformly 54 in this case study. If we had used the 2-level column, namely column 1, or if we had used the dummy level technique (described in Chapter 7), the entry in category (V) would not be 54. The probabilities for the cumulative categories shown in Table 5.5 are obtained by dividing the number of observations in each cumulative category by the entry in the last cumulative category for that factor level, which is 54 for the present case.

TABLE 5.4 CATEGORIZED DATA FOR SURFACE DEFECTS

Expt. No.	Number of Observations by Categories					Number of Observations by Cumulative Categories				
	I	II	III	IV	V	(I)	(II)	(III)	(IV)	(V)
1	9	0	0	0	0	9	9	9	9	9
2	5	2	2	0	0	5	7	9	9	9
3	1	0	6	2	0	1	1	7	9	9
4	0	8	1	0	0	0	8	9	9	9
5	0	1	0	4	4	0	1	1	5	9
6	1	0	4	1	3	1	1	5	6	9
7	0	1	1	4	3	0	1	2	6	9
8	3	0	2	1	3	3	3	5	6	9
9	0	0	0	4	5	0	0	0	4	9
10	9	0	0	0	0	9	9	9	9	9
11	8	1	0	0	0	8	9	9	9	9
12	2	3	3	0	1	2	5	8	8	9
13	4	2	2	1	0	4	6	8	9	9
14	2	3	4	0	0	2	5	9	9	9
15	0	1	1	1	6	0	1	2	3	9
16	3	4	2	0	0	3	7	9	9	9
17	2	1	0	2	4	2	3	3	5	9
18	0	0	0	2	7	0	0	0	2	9
Total	49	27	28	22	36	49	76	104	126	162

The third step in data analysis is to plot the cumulative probabilities. Two useful plotting methods are the line plots shown in Figure 5.5 and the bar plots shown in Figure 5.6. From both figures, it is apparent that temperature (factor A) and pressure (factor B) have the largest impact on the cumulative distribution function for the surface defects. The effects of the remaining four factors are small compared to temperature and pressure. Among the factors C, D, E, and F, factor F has a somewhat larger effect.

In the line plots of Figure 5.5, for each control factor we look for a level for which the curve is uniformly higher than the curves for the other levels of that factor.

TABLE 5.5 FACTOR EFFECTS FOR THE CATEGORIZED SURFACE DEFECT DATA

Factor	Level	Number of Observations by Cumulative Categories					Probabilities for the Cumulative Categories				
		(I)	(II)	(III)	(IV)	(V)	(I)	(II)	(III)	(IV)	(V)
A. Temperature (°C)	$A_1: T_0-25$	34	40	51	53	54	0.63	0.74	0.94	0.98	1.00
	$A_2: T_0$	7	22	34	41	54	0.13	0.41	0.63	0.76	1.00
	$\overline{A_3: T_0+25}$	8	14	19	32	54	0.15	0.26	0.35	0.59	1.00
B. Pressure (mtorr)	$B_1: P_0-200$	25	40	46	51	54	0.46	0.74	0.85	0.94	1.00
	$B_2: P_0$	20	28	36	43	54	0.37	0.52	0.67	0.80	1.00
	$\overline{B_3: P_0+200}$	4	8	22	32	54	0.07	0.15	0.41	0.59	1.00
C. Nitrogen (sccm)	$C_1: N_0$	19	30	32	39	54	0.35	0.56	0.59	0.72	1.00
	$\overline{C_2: N_0-150}$	11	20	28	39	54	0.20	0.37	0.52	0.72	1.00
	$C_3: N_0-75$	19	26	44	48	54	0.35	0.48	0.81	0.89	1.00
D. Silane (sccm)	$D_1: S_0-100$	20	25	34	41	54	0.37	0.46	0.63	0.76	1.00
	$D_2: S_0-50$	13	31	42	44	54	0.24	0.57	0.78	0.81	1.00
	$D_3: S_0$	16	20	28	41	54	0.30	0.37	0.52	0.76	1.00
E. Settling time (min)	$E_1: t_0$	21	27	38	43	54	0.39	0.50	0.70	0.80	1.00
	$\overline{E_2: t_0+8}$	16	29	36	42	54	0.30	0.54	0.67	0.78	1.00
	$E_3: t_0+16$	12	20	30	41	54	0.22	0.37	0.56	0.76	1.00
F. Cleaning method	$F_1: None$	21	23	26	34	54	0.39	0.43	0.48	0.63	1.00
	$\overline{F_2: CM_2}$	21	30	40	46	54	0.39	0.56	0.74	0.85	1.00
	$F_3: CM_3$	7	23	38	46	54	0.13	0.43	0.70	0.85	1.00

A uniformly higher curve implies that the particular factor level produces more observations with lower defect counts; hence, it is the best level. In Figure 5.6, we look for a larger height of category I and smaller height of category V. From the two figures, it is clear that A_1, B_1, and F_2 are the best levels for the respective factors. The choice of the best level is not as clear for the remaining three factors. However, the curves for the factor levels C_2, D_3, and E_3 lie uniformly lower among the curves for all levels of the respective factors, and these levels must be avoided. Thus, the optimum settings suggested by the analysis are $A_1B_1(C_1/C_3)(D_1/D_2)(E_1/E_2)F_2$. By comparing Figures 5.5, 5.6, and 4.5 it is apparent that the conclusions based on the ordered categorical data are consistent with the conclusions based on the actual counts, except for factors C, D, and E whose effects are rather small.

The next step in the analysis is to predict the distribution of defect counts under the starting and optimum conditions. This can be achieved analogous to the procedure described in Chapter 3, except that we must use the omega transform, also known as

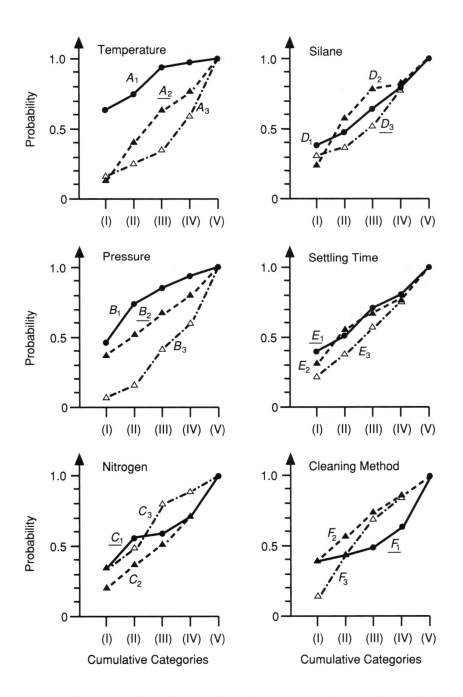

Figure 5.5   Line plots of the factor effects for the categorized surface defect data.

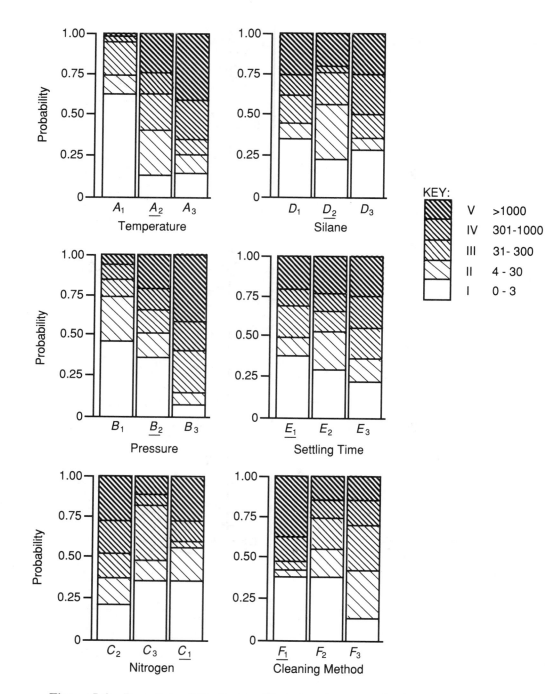

Figure 5.6   Bar plots of the factor effects for the categorized surface defect data.

the logit transform, of the probabilities for the cumulative categories (see Taguchi [T1]). The omega transform for probability p is given by the following equation:

$$\omega(p) = 10 \log_{10} \frac{p}{1-p}.$$

Note that the omega transform is the same as the S/N ratio for the fraction defective type of static problems.

Let us take the optimum settings recommended in Chapter 4, namely $A_1B_2C_1D_3E_2F_2$, to illustrate the computation of the predicted distribution. Since, according to Figure 5.5, factors C, D, and E have a small effect, we will only use the effects of factors A, B, and F in the prediction formula.

The average probability for category (I) taken over 18 experiments is $\mu_{(I)} = 49/162 = 0.30$ (see Table 5.4). Referring to Table 5.5, the predicted omega value for category (I), denoted by $\omega_{A_1B_2C_1D_3E_2F_2(I)}$, can be computed as follows:

$$\omega_{A_1B_2C_1D_3E_2F_2(I)} = \omega_{\mu(I)} + \left[\omega_{A_1(I)} - \omega_{\mu(I)}\right]$$

$$+ \left[\omega_{B_2(I)} - \omega_{\mu(I)}\right] + \left[\omega_{F_2(I)} - \mu_{\mu(I)}\right]$$

$$= \omega(0.30) + [\omega(0.63) - \omega(0.30)]$$

$$+ [\omega(0.37) - \omega(0.30)] + [\omega(0.39) - \omega(0.30)]$$

$$= -3.68 + [2.31 + 3.68] + [-2.31 + 3.68]$$

$$+ [-1.94 + 3.68]$$

$$= 5.42 \text{ dB} .$$

Then, by the inverse omega transform, the predicted probability for category (I) is 0.78. Predicted probabilities for the cumulative categories (II), (III) and (IV) can be obtained analogously. Prediction is obviously 1.0 for category (V). The predicted probabilities for the cumulative categories for the starting and the optimum settings are listed in

Table 5.6. These probabilities are also plotted in Figure 5.7. It is clear that the recommended optimum conditions give much higher probabilities for the low defect count categories when compared to the starting conditions. The probability of 0-3 defects, (category I), is predicted to increase from 0.23 to 0.78 by changing from starting to the optimum conditions. Likewise, the probability for the 1001 and more category reduces from 0.37 to 0.01.

The predicted probabilities for the cumulative categories should be compared with those observed under the starting and optimum conditions to verify that the additive model is appropriate for the case study. This is the same as the Step 8 of the Robust Design cycle described in Chapter 4. Selection of appropriate orthogonal arrays for a case study, as well as the confounding of 2-factor interactions with the main effects, is discussed in Chapter 7. However, we note here that when accumulation analysis is used, 3-level orthogonal arrays should be preferred over 2-level orthogonal arrays for minimizing the possibility of misleading conclusions about the factor effects. Particularly, the orthogonal arrays L_{18} and L_{36} are most suitable. In any case, the verification experiment is important for ensuring that the conclusions about the factor effects are valid.

5.6 SUMMARY

- The quadratic loss function is ideally suited for evaluating the quality level of a product as it is shipped by a supplier to a customer. It typically has two components: one related to the deviation of the product's function from the target, and the other related to the sensitivity to noise factors.

- S/N ratio developed by Genichi Taguchi, is a predictor of quality loss after making certain simple adjustments to the product's function. It isolates the sensitivity of the product's function to noise factors. In Robust Design we use the S/N ratio as the objective function to be maximized.

- Benefits of using the S/N ratio for optimizing a product or process design are:

  — Optimization does not depend on the target mean function. Thus, the design can be reused in other applications where the target is different.

  — Design of subsystems and components can proceed in parallel. Specifications for the mean function of the subsystems and components can be changed later during design integration without adversely affecting the sensitivity to noise factors.

  — Additivity of the factor effects is good when an appropriate S/N ratio is used. Otherwise, large interactions among the control factors may occur, resulting in high cost of experimentation and potentially unreliable results.

TABLE 5.6 PREDICTED PROBABILITIES FOR THE CUMULATIVE CATEGORIES

Control Factor Settings	ω Values for the Cumulative Categories					Probabilities for the Cumulative Categories				
	(I)	**(II)**	**(III)**	**(IV)**	**(V)**	**(I)**	**(II)**	**(III)**	**(IV)**	**(V)**
Optimum $A_1B_2C_1D_3E_2F_2$	5.42	6.98	14.53	19.45	∞	0.78	0.83	0.97	0.99	1.00
Starting $A_2B_2C_1D_3E_1F_1$	−3.68	−1.41	0.04	2.34	∞	0.23	0.42	0.50	0.63	1.00

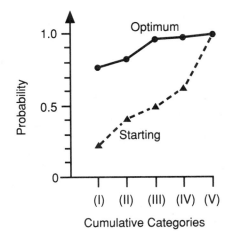

Figure 5.7   Predicted probabilities for the cumulative categories.

— Overall productivity of the development activity is improved.

- Robust Design problems can be divided into two broad classes: static problems, where the target value for the quality characteristic is fixed, and dynamic problems, where the quality characteristic is expected to follow the signal factor.

- Common types of static problems and the corresponding S/N ratios are summarized in Table 5.7.

- Common types of dynamic problems and the corresponding S/N ratios are summarized in Table 5.8.

- For the problems where an adjustment factor does not exist, the optimization is done by simply maximizing the S/N ratio.

- For the problems where an adjustment factor exists, the problem can be generically stated as minimize sensitivity to noise factors while keeping the mean function on target. By using S/N ratio, these problems can be converted into unconstrained optimization problems and solved by the following two-step procedure:

  1. Maximize S/N ratio, without worrying about the mean function.

  2. Adjust the mean on target by using the adjustment factor.

- The optimization strategy consists of the following steps:

  1. Evaluate the effects of control factors under consideration on η and on the mean function.

  2. For factors that have a significant effect on η, select levels that maximize η.

  3. Select any factor that has no effect on η but a significant effect on the mean function as an adjustment factor. Use it to bring the mean function on target. (In practice, we must sometimes settle for a factor that has a small effect on η but a significant effect on the mean function as an adjustment factor.)

  4. For factors that have no effect on η and the mean function, we can choose any level that is convenient from other considerations such as cost or other quality characteristics.

TABLE 5.7 S/N RATIOS FOR STATIC PROBLEMS

Problem Type	Range for the Observations	Ideal Value	Adjustment	S/N Ratio and Comments
Smaller-the-better type	$0 \leq y < \infty$	0	None	$\eta = -10 \log_{10} \left[\dfrac{1}{n} \sum\limits_{i=1}^{n} y_i^2 \right]$
Nominal-the-best type	$0 \leq y < \infty$	Nonzero, finite	Scaling	$\eta = 10 \log_{10} \dfrac{\mu^2}{\sigma^2}$ $\mu = \dfrac{1}{n} \sum\limits_{i=1}^{n} y_i$ $\sigma^2 = \dfrac{1}{n-1} \sum\limits_{i=1}^{n} (y_i - \mu)^2$
Larger-the-better type	$0 \leq y < \infty$	∞	None	$\eta = -10 \log_{10} \left[\dfrac{1}{n} \sum\limits_{i=1}^{n} \dfrac{1}{y_i^2} \right]$
Signed-target	$-\infty < y < \infty$	Finite, usually 0	Leveling	$\eta = -10 \log_{10} \sigma^2$ $\sigma^2 = \dfrac{1}{n-1} \sum\limits_{i=1}^{n} (y_i - \mu)^2$
Fraction defective	$0 \leq p \leq 1$	0	None	$\eta = -10 \log_{10} \left[\dfrac{p}{1-p} \right]$
Ordered categorical				Use accumulation analysis. See Section 5.5.
Curve or vector response				Divide the problem into several individual problems of the above types. See Chapter 6.

TABLE 5.8 S/N RATIOS FOR DYNAMIC PROBLEMS

Problem Type	Input Range	Output Range	Ideal Function	Adjust-ment	S/N Ratio and Comments
Continuous-continuous (C-C)	$-\infty < M < \infty$	$-\infty < y < \infty$; $y = 0$ when $M = 0$	$y = M$	Scaling	$\eta = 10 \log_{10} \dfrac{\beta^2}{\sigma_e^2}$ $$\beta = \dfrac{\sum\limits_{i=1}^{m}\sum\limits_{j=1}^{n}(y_{ij}M_i)}{\sum\limits_{i=1}^{m}\sum\limits_{j=1}^{n}(M_i^2)}$$ $$\sigma_e^2 = \dfrac{1}{(mn-1)}\sum_{i=1}^{m}\sum_{j=1}^{n}(y_{ij}-\beta M_i)^2$$
Continuous-digital (C-D)	$-\infty < M < \infty$	Binary; ON, OFF			Divide the problem into two separate problems of C-C or nominal-the-best type for the ON and OFF functions. See Chapter 9.
Digital-continuous (D-C)	Binary; 0, 1	$-\infty < y < \infty$			Divide the problem into two nominal-the-best type problems.
Digital-digital (D-D)	Binary; 0, 1	Binary; 0, 1	$y = M$	Leveling	$\eta = 10 \log_{10}\left[\dfrac{(1-2p')^2}{p'(1-p')}\right]$ $$p' = \left[1 + \sqrt{\dfrac{1-p}{p}\cdot\dfrac{1-q}{q}}\right]^{-1} =$$ equalized error probability p = error probability of output being 1 when input is 0 q = error probability of output being 0 input is 1

Chapter 6

ACHIEVING ADDITIVITY

The goal of a Robust Design project is to determine the best levels of each of the various control factors in the design of a product or a manufacturing process. To select the best levels (or settings) of the control factors, we must first be able to predict the product's performance and robustness for any combination of the settings. Further, the prediction must be valid not only under the laboratory conditions but also under manufacturing and customer conditions.

As pointed out in Chapter 3, if the effects of the control factors on performance and robustness are additive (that is, they follow the superposition principle), then we can predict the product's performance for any combination of levels of the control factors by knowing only the main effects of the control factors. The experimental effort needed for estimating these effects is rather small. On the other hand, if the effects are not additive (that is, interactions among the control factors are strong), then we must conduct experiments under all combinations of control factor settings to determine the best combination. This is clearly expensive, especially when the number of control factors is large.

A more important reason exists for seeking additivity. The conditions under which experiments are conducted can also be considered as a control factor. There are three types of these conditions: laboratory, manufacturing, and customer usage. If strong interactions among the control factors are observed during laboratory experiments, these control factors are also likely to interact with conditions of experimentation. Consequently, the levels found optimum in the laboratory may not prove to be optimum under manufacturing or customer conditions. Thus, the manufacturing reject or rework rate may turn out to be high; costly design changes may become necessary; the product may fail in the field sooner than expected; and the product may not function on target under different customer environments.

Indeed, the presence of large interaction among control factors is an indication of later problems. The interactions that are particularly harmful are those where the direction of a control factor's effect on performance or robustness changes when the level of another control factor is changed. Interactions that are less harmful include those where the direction of a factor's effect does not change, but, rather, the magnitude of the effect changes when the level of another factor is changed. Thus, during Robust Design experiments, we should not only find the best settings of the control factors but also try to achieve additivity of factor effects. Orthogonal array experiments, followed by verification experiments, provide a powerful tool for economizing on the number of experiments needed for studying many control factors simultaneously. They also provide a powerful tool for evaluating the additivity of factor effects.

Although additivity cannot be guaranteed, the relative magnitude and importance of interactions can be reduced greatly through the proper choice of quality characteristics, objective functions [the signal-to-noise (S/N) ratios], and control factors. Recall that a quality characteristic is a characteristic we use to measure the product's quality in a Robust Design case study.

Selecting the quality characteristic, testing conditions for evaluating sensitivity to noise factors, S/N ratio, and control factors are difficult tasks that usually constitute the bulk of the effort in planning Robust Design case studies. It requires the engineering know-how of the specific project and also knowledge of the Robust Design method. Care taken in this activity can greatly enhance the ability of the Robust Design experiment to generate dependable and reproducible information with a small number of experiments. This chapter describes the important considerations in selecting quality characteristics, S/N ratios, and control factors and their levels. This chapter has six sections:

- Section 6.1 describes guidelines for selecting quality characteristics.

- Section 6.2 gives examples of quality characteristics.

- Section 6.3 describes the process of selecting the S/N ratio with illustrative examples.

- Section 6.4 discusses the selection of control factors and their levels.

- Section 6.5 describes the role of orthogonal arrays.

- Section 6.6 summarizes the important points of this chapter.

The discussion in this chapter is based on a paper by Phadke and Taguchi [P7].

6.1 GUIDELINES FOR SELECTING QUALITY CHARACTERISTICS

In designing a product, one is invariably interested in increasing the mean time to failure, whereas in designing a manufacturing process one wants to maximize the yield. The final success of the product or the process depends on how well such responses (reliability, yield, etc.) meet the customer's expectations. However, as illustrated in Sections 6.2 and 6.3, such responses are not necessarily suitable as quality characteristics for optimizing the product design. The following guidelines are useful in selecting the quality characteristics. The concepts behind these guidelines are new and will become clear through the examples presented in the following sections. To select quality characteristics, you should do the following:

1. Identify the ideal function or the ideal input-output relationship for the product or the process. The quality characteristic should be directly related to the *energy transfer* associated with the basic mechanism of the product or the process.

2. As far as possible, choose *continuous variables* as quality characteristics.

3. The quality characteristics should be *monotonic*—that is, the effect of each control factor on robustness should be in a consistent direction, even when the settings of other control factors are changed. In several situations, it is difficult to judge the monotonicity of a quality characteristic before conducting experiments. In such situations, matrix experiments followed by confirmation experiments are the only way of determining if the quality characteristics have monotonicity.

4. Try to use quality characteristics that are *easy to measure*. Availability of appropriate measurement techniques often plays an important role in the selection of quality characteristics.

5. Ensure that quality characteristics are *complete*—that is, they should cover all dimensions of the ideal function or the input-output relationship.

6. When a product has a built-in feedback mechanism, *optimize the open loop, the sensor, and the compensation modules separately,* and then integrate the three modules. Similarly, complex products should be divided into convenient modules. Each module should be optimized separately and the modules should then be integrated together. While optimizing a particular module, the variation in other modules should be treated as noise factors. This is important for smooth system integration.

Finding quality characteristics that meet all of these guidelines is sometimes difficult or simply not possible with the technical know-how of the engineers involved. However, the Robust Design experiment will be inefficient to the extent these guidelines are not satisfied.

While conducting Robust Design experiments for manufacturing processes, appropriate productivity characteristics (such as processing time) should be measured in addition to quality characteristics so that appropriate economic trade-offs can also be made in selecting the best levels of the control factors.

6.2 EXAMPLES OF QUALITY CHARACTERISTICS

Different types of variables can be used as quality characteristics. For example, a quality characteristic could be the output or the response variable. It could also be the threshold value of a suitable control factor, signal factor, or noise factor for achieving a certain value of the output. Typically, when the output is discrete, such as ON-OFF states, it becomes necessary to use the threshold values. (See the paper feeder example in Section 6.3 and the temperature control circuit design in Chapter 9.) But this is not the only case where control, signal, or noise factors are used as quality characteristics.

This section presents some examples of quality characteristics, analyzes how well they meet the guidelines of the previous section, and examines the impact on experimental efficiency.

Yield as a Quality Characteristic

Many Robust Design experiments are aimed at improving the yield of manufacturing processes. However, if yield is used as the quality characteristic in these experiments, it is possible to lose monotonicity, which will lead to an unnecessarily large number of experiments. As an example, consider a photolithography process used in integrated circuit manufacturing to print lines of a certain width. The percentage of microchips with line widths within the limits 2.75 to 3.25 micrometers represents the yield of good microchips. This is the customer-observable response we want to maximize. Let us examine the problem of using this response as the quality characteristic.

Exposure and develop time are two important control factors for this photolithography process. For certain settings of these factors, called initial levels, the yield is 40 percent as indicated in Table 6.1. When the exposure time alone is increased to its high setting, the yield becomes 75 percent. Also, when the develop time alone is increased to its high setting, the yield increases to 75 percent. Thus, we may anticipate that increasing both the exposure time and the develop time would improve the yield beyond 75 percent. But when that is done, the yield drops down to 40 percent. This is the lack of monotonicity. Interactions among control factors are critically important when there is lack of monotonicity. In such situations, we need to study all combinations of control factors to find the best settings. With only two control factors, studying all combinations is not a major issue. But with eight or ten control factors, the experimental resources needed would be prohibitively large.

What is, then, a better quality characteristic for the photolithography process? To answer this question, let us look at Table 6.2 which shows not only the yield (that is, the percentage of chips with the desired line width) but also the percentage of chips with line widths smaller or larger than the desired line width. Such data are called ordered categorical data. The reason for getting low yield, when both the exposure and the develop time are set at high levels, becomes clear from this table: the effect of each of these two control factors is to increase the overall line width. Consequently, recording the data in all three categories (small, desired, and large) provides a better quality characteristic than only yield by itself. The monotonicity can be observed through the cumulative categories—small, and small plus desired—as shown in Table 6.2.

TABLE 6.1 OBSERVED YIELD FOR DIFFERENT EXPOSURE AND DEVELOP TIMES

Expt. No.	Exposure	Develop Time	Yield (%)*
1	Initial	Initial	40
2	High	Initial	75
3	Initial	High	75
4	High	High	40

* Percent of chips with line width in the
range 2.75 to 3.25 micrometers.

TABLE 6.2 ORDERED CATEGORICAL DATA ON LINE WIDTH

Expt. No.	Exposure	Develop Time	Line Width			Cumulative Line Width		
			Small* (%)	Desired† (%)	Large‡ (%)	Small (%)	Small + Desired (%)	Small + Desired + Large (%)
1	Initial	Initial	55	40	5	55	95	100
2	High	Initial	10	75	15	10	85	100
3	Initial	High	10	75	15	10	85	100
4	High	High	3	40	57	3	43	100

* *Small* means line width <2.75 micrometers.
† *Desired* means line width is in the range 2.75 to 3.25 micrometers.
‡ *Large* means line width >3.25 micrometers.

The proper quality characteristic for photolithography is the actual line-width measurement, for example, 2.8 or 3.1 micrometers. When the line-width distribution is known, it is an easy task to compute the yield. Note that the line width is directly related to the amount of energy transferred during exposure and developing. The more the energy transferred, the larger is the line width. Also, our experience with photolithography has shown that the actual line width measurement is a monotonic characteristic with respect to the control factors. Further, when the target dimensions are changed, we can use the same experimental data to determine the new optimum settings of the control factor. In this case, the measurement technique is not much of an issue, although taking categorical data (small, desired, or large) is generally a little easier than recording actual measurements. A case study where the actual line width was used to optimize a photolithography process used in VLSI circuit fabrication is given by Phadke, Kackar, Speeney and Grieco [P5].

In summary, for the photolithography example, yield is the worst quality characteristic, ordered categorical data are better, and actual line width measurement (continuous variable) is the best.

Spray Painting Process

This example vividly illustrates the importance of energy transfer in selecting a quality characteristic. Sagging is a common defect in spray painting. It is caused by formation of large paint drops that flow downward due to gravity. Is the distance through which the paint drops sag a good quality characteristic? No, because this distance is primarily controlled by gravity, and it is not related to the basic energy transfer in spray painting. However, the size of the drops created by spray painting is directly related to energy transfer and, thus, is a better quality characteristic. By taking the size of the drops as the quality characteristic, we can block out the effect of gravity, an extraneous phenomenon for the spray painting process.

Yield of a Chemical Process

There are many chemical processes that begin with a chemical A, which after reaction, becomes chemical B and, if the reaction is allowed to continue, turns into chemical C. If B is the desired product of the chemical process, then considering the yield of B as a quality characteristic is a poor choice. As in the case of photolithography, the yield is not a monotonic characteristic. A better quality characteristic for this experiment is the concentration of each of the three chemicals. The concentration of A and the concentration of A plus B possess the needed monotonicity property.

6.3 EXAMPLES OF S/N RATIOS

Basic types of Robust Design problems and the associated S/N ratios were described in Chapter 5. A majority of Robust Design projects fall into one of these basic types of problems. This section gives three examples to illustrate the process of classification of Robust Design problems. Two of these examples also show how a complex problem can be broken down into a composite of several basic types of problems.

Heat Exchanger Design

Heat exchangers are used to heat or cool fluids. For example, in a refrigerator a heat exchanger coil is used inside the refrigerator compartment to transfer the heat from the air in the compartment to the refrigerant fluid. This leads to lowering of the temperature inside the compartment. Outside the refrigerator, the heat from the refrigerant is transferred to the room air through another heat exchanger.

In optimizing the designs of heat exchangers and other heat-transfer equipment, defining the reference temperature is critical so that the optimization problem can be correctly classified.

Consider the heat exchanger shown in Figure 6.1, which is used to cool the fluid inside the inner tube. The inlet temperature of the fluid to be cooled is T_1. As the fluid moves through the tube, it loses heat progressively to the fluid outside the tube; its outlet temperature is T_2. The inlet and outlet temperature for the coolant fluid are T_3 and T_4, respectively. Let the target outlet temperature for the fluid being cooled be T_0. Also, suppose the customer's requirement is that $|T_2 - T_0| < 10$ °C. What is the correct quality characteristic and S/N ratio for this Robust Design problem?

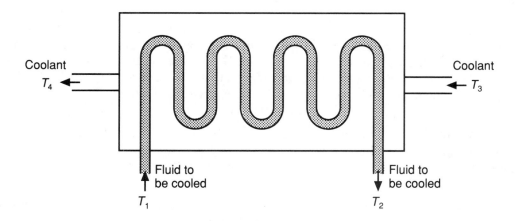

Figure 6.1   Schematic diagram of a heat exchanger.

One choice is to take target temperature T_0 as the reference temperature and $y = |T_2 - T_0|$ as the quality characteristic. Minimizing y is then the goal of this experiment; thus, at first it would appear that y should be treated as a smaller-the-better characteristic where the mean square value of y is minimized. The difficulty with this formulation of the problem is that by taking the square of y the positive and negative deviations in temperature are treated similarly. Consequently, interactions become important. This can be understood as follows: If y is too large because T_2 is too large compared to T_0, then y can be reduced by increasing the coil length. Note that a longer coil length leads to more cooling of the fluid and, hence, smaller T_2. On the contrary, if y is too large because T_2 is too small, then y can be reduced by decreasing the coil length. Thus, there are two opposite actions that can reduce y, but they cannot be distinguished by observing y. Therefore, y is not a good quality characteristic, and this problem should not be treated as smaller-the-better type.

Here, the proper reference temperature is T_3 because it represents the lowest temperature that could be achieved by the fluid inside the tube. Thus, the correct quality characteristic is $y' = T_2 - T_3$. Note that y' is always positive. Also, when the mean of y' is zero, its variance must also be zero. Hence, the problem should be classified as a nominal-the-best type with the target value of y' equal to $T_0 - T_3$. This formulation does not have the complication of interaction we described with y as the quality characteristic. Furthermore, if the target temperature T_0 were changed, the information obtained using y' as the quality characteristic would still be useful. All that is necessary is to adjust the mean temperature on the new target. However, if y were used as the quality characteristic, the design would have to be reoptimized when T_0 is changed, which is undesirable. This loss of reusability is one of the reasons for lower R&D productivity.

Paper Handling in Copying Machines

In a copying machine, a critical customer-observable response is the number of pages copied before a paper-handling failure. In designing the paper-handling system we might take λ, the number of pages copied before failure, as the quality characteristic. However, in this case, the number of pages that would have to be copied during the copier development would be excessively large. Also, decoupling the designs of the various modules is not possible when λ is taken as the quality characteristic.

A close look at the paper-handling equipment reveals that there are two basic functions in paper handling: paper feeding and paper transport. Paper feeding means picking a sheet, either the original or a blank sheet. Paper transport means moving the sheet from one station to another.

A schematic diagram of a paper feeder is shown in Figure 6.2(a). The two main defects that arise in paper feeding are: no sheet fed or multiple sheets fed. A fundamental characteristic that controls paper feeding is the normal force needed to pick up a sheet. Thus, we can measure the threshold force, F_1, to pick up just one sheet and the threshold force, F_2, to pick up two sheets. Note that the normal force is a control factor and that F_1 and F_2 meet the guidelines listed in Section 6.1 and are better quality characteristics compared to λ. By making F_1 as small as possible and F_2 as large as possible, we can improve the operating window $F_1 - F_2$ [see Figure 6.2(b)], reduce both types of paper feeding defects, and thus increase λ. The idea of enlarging the operating window as a means of improving product reliability is due to Clausing [C2].

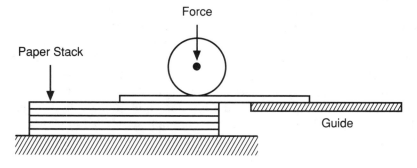

(a) Schematic Diagram of a Paper Feeder

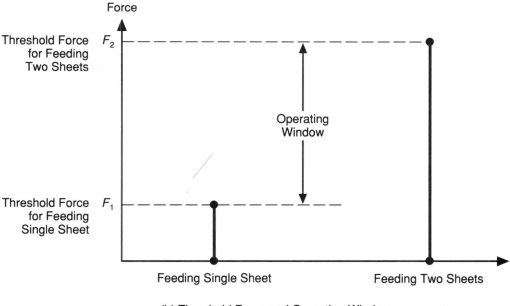

(b) Threshold Force and Operating Window

Figure 6.2   Design of a paper feeder.

Here, the appropriate S/N ratios for F_1 and F_2 are, respectively, the smaller-the-better type and the larger-the-better type. Note that the two threshold forces comprise a *vector quality characteristic*. We must measure and optimize both of them. This is what we mean by *completeness* of a quality characteristic.

In a copying machine, paper is moved through a 3-dimensional path using several paper transport modules. Figure 6.3 shows a planar diagram for paper movement through a single module. The fundamental characteristics in transporting paper are the (x,y) movements of the center of the leading edge of the paper, the rotation angle θ of the leading edge, and the time of arrival of the paper at the next module. The lateral movement (y movement) of the paper can be taken care of by registration against a guide. The remaining characteristics can then be addressed by placing two sensors to measure the arrival times of the left and right ends of the leading edge of the paper. Both of these arrival times have a common nonzero target mean and can be classified as nominal-the-best type problems.

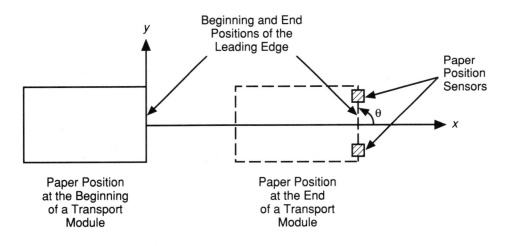

Figure 6.3   Planar diagram for paper transport.

Here, also, the two arrival times can be viewed as a vector quality characteristic. Both times must be measured and optimized. If we omit one or the other, we cannot guard against failure due to the paper getting twisted during transport. Also, by optimizing for both the arrival times (that is, minimizing the variability of both the arrival times and making their averages equal to each other), the design of each paper transport module can be decoupled from other modules. Optimizing each of the paper-feeding and paper-transport characteristics, described above, automatically optimizes λ. Thus, the problem of life improvement is broken down into several problems of nominal-the-best, smaller-the-better, and larger-the-better types. It is quite obvious that optimizing these separate problems automatically improves λ, the number of pages copied before failure.

Electrical Filter Design

Electrical filters are used widely in many electronic products, including telecommunications and audio/video equipment. These circuits amplify (or attenuate) the components of the input voltage signal at different frequencies according to the specified frequency response function [see Figure 6.4(a)]. An equalizer, used in high-fidelity audio equipment, is an example of a filter familiar to many people. It is used to amplify or attenuate the sounds of different musical instruments in a symphony orchestra.

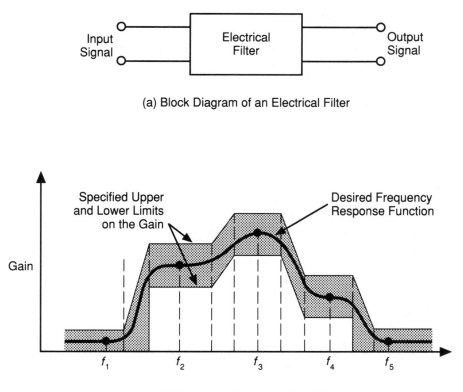

(a) Block Diagram of an Electrical Filter

(b) Frequency Response Function

Figure 6.4   Design of an electrical filter.

Figure 6.4(b) shows an example of a desired frequency response function and the customer-specified upper and lower limits for the gain. If the customer-specified gain limits are violated at any frequency, the filter is considered defective. From the preceding discussion in this chapter, it should be apparent that counting the percentage of defective filters, though easiest to measure, is not a good quality characteristic.

This problem can be solved more efficiently by dividing the frequencies into several bands, say five bands as shown in Figure 6.4(b). For each of the middle three bands, we must achieve gain equal to the gain specified by the frequency response function. Therefore, we treat these as three separate nominal-the-best type problems. For each band, we must identify a separate adjustment factor that can be used to set the mean gain at the right level. Note that a resistor, capacitor, or some other component in the circuit can serve as an adjustment factor. For any one of these bands, the adjustment factors for the other two bands should be included as noise factors, along with other noise factors, such as component tolerances and temperature. Then, adjusting the gain in one band would have a minimal effect on the mean gain in the other bands.

For each of the two end bands, we must make the gain as small as possible. Accordingly, these two bands belong to the class of smaller-the-better type problems. Thus, we have divided a problem where we had to achieve a desired curvilinear response into several familiar problems.

6.4 SELECTION OF CONTROL FACTORS

Additivity of the effects of the control factors is also influenced by the selection of the control factors and their levels. By definition, the control factors are factors whose levels can be selected by the designer. Next, it is important that each control factor influence a distinct aspect of the basic phenomenon affecting the quality characteristic. If two or more control factors affect the same aspect of the basic phenomenon, then the possibility of interaction among these factors becomes high. When such a situation is recognized, we can reduce or even eliminate the interaction through proper transformation of the control factor levels. We refer to this transformation as *sliding levels*. The following examples illustrate some of the important considerations in the selection of control factors. A qualitative understanding of how control factors affect a product is very important in their selection.

Treatment of Asthmatic Patients

This rather simplified example brings out an important consideration in the selection of control factors. Consider three drugs (A, B, and C) proposed by three scientists for treating wheezing in asthmatic patients. Suppose the drug test results indicate that if no drug is given, the patient's condition is bad. If only drug A is given, the patients get somewhat better; if only drug B is given, the patients feel well; and if only drug C is given, the patients feel moderately good. Can the three drugs be considered as three separate control factors? If so, then a natural expectation is that by giving all three drugs simultaneously, we can make the patients very well.

Suppose we take a close look at these drugs to find out that all three drugs contain theophillin as an active ingredient, which helps dilate the bronchial tubes. Drug A has 70 percent of full dose, drug B has 100 percent of full dose, and drug C has 150 percent of full dose. Administering all three drugs simultaneously implies giving 320 percent of full dose of theophillin. This could significantly worsen the patient's condition. Therefore, the three drugs interact. The proper way to approach this problem is to think of the theophillin concentration as a single control factor with four levels: 0 percent (no drug), 70 percent (drug A), 100 percent (drug B), and 150 percent (drug C). Here the other ingredients of the three drugs should be examined as additional potential control factors.

Photolithography Process

Aperture and exposure are among the two important control factors in the photolithography process used in VLSI fabrication (see Phadke, Kackar, Speeney, and Grieco [P5]). The width of the lines printed by photolithography depends on the depth of field and the total light energy falling on the photoresist. The aperture alone determines the depth of field. However, both aperture and exposure time influence the total light energy. In fact, the total light energy for fixed light intensity is proportional to the product of the aperture and exposure time. Thus, if we chose aperture and exposure time as control factors, we would expect to see strong interaction between these two factors. The appropriate control factors for this situation are aperture and total light energy.

Suppose 1.2N, N, and 0.8N are used as three levels for light energy, where N stands for the nominal level or the middle level. We can achieve these levels of light energy for various apertures through the sliding levels of exposure as indicated in Table 6.3. The level N of total light energy can be achieved by setting exposure at 120 when aperture is 1, exposure at 90 when aperture is 2, and exposure at 50 when aperture is 3.

TABLE 6.3 SLIDING LEVELS FOR EXPOSURE IN PHOTOLITHOGRAPHY PROCESS

		Exposure (PEP-Setting)		
		1.2N	N	0.8N
	1	96	120	144
Aperture	2	72	90	108
	3	40	50	60

The thickness of the photoresist layer is another fundamental characteristic that influences the line width. The thickness can be changed by controlling the photoresist viscosity and the spin speed. Here too, sliding levels of spin speed should be considered to minimize interactions (see Phadke, Kackar, Speeney, and Grieco [P5]).

6.5 ROLE OF ORTHOGONAL ARRAYS

Matrix experiments using orthogonal arrays play a crucial role in achieving additivity—they provide a test to see whether interactions are large compared to the main effects.

Consider a matrix experiment where we assign only main effects to the columns of an orthogonal array so that the interactions (2-factor, 3-factor, etc.) are confounded with the main effects (see Chapter 7). There are two possibilities for the relative magnitudes of the interactions:

1. If one or more of these interactions are large compared to the main effects, then the main effects with which these interactions are confounded will be estimated with large bias or error. Consequently, the observed response under the predicted optimum conditions will not match the prediction based on the additive model. Thus, in this case the verification experiment will point out that large interactions are present.

2. On the contrary, if the interactions are small compared to the main effects, then the observed response under the predicted optimum conditions will match the prediction based on the additive model. Thus, in this case the verification experiment will confirm that the main effects dominate the interactions.

Optimization studies where only one factor is studied at a time are not capable of determining if interactions are or are not large compared to the main effects. Thus, it is important to conduct multifactor experiments using orthogonal arrays. Dr. Taguchi considers the ability to detect the presence of interactions to be the primary reason for using orthogonal arrays to conduct matrix experiments.

Sections 6.2, 6.3, and 6.4 described the engineering considerations in selecting the quality characteristics, S/N ratios, and control factors and their levels. Matrix experiments using orthogonal arrays provide a test to see whether the above selections can successfully achieve additivity. If additivity is indeed achieved, the matrix experiment provides simultaneously the optimum values for the control factors. If additivity is not achieved, the matrix experiment points it out so that one can re-examine the selection of the quality characteristics, S/N ratios, and control factors and their levels.

6.6 SUMMARY

• Ability to predict the robustness (sensitivity to noise factors) of a product for any combination of control factor settings is needed so that the best control factor levels can be selected. The prediction must be valid, not only under the laboratory conditions, but also under manufacturing and customer usage conditions.

- It is important to have additivity of the effects of the control factors on the sensitivity to noise factors (robustness) for the following reasons:

  — Only main effects need to be estimated which takes only a small number of experiments. However, if the interactions among the control factors are strong, experiments must be conducted under all combinations of control factor settings, which is clearly expensive, if not impractical.

  — Conditions under which experiments are conducted can also be considered as a control factor. The conditions consist of three types: laboratory, manufacturing, and customer usage. Presence of strong interactions among the control factors studied in a laboratory is an indication that the experimental conditions are likely to interact with the control factors that have been studied. This interaction, if present, can make the laboratory results invalid, which leads to product failure during manufacturing and customer usage.

- The additivity is influenced greatly by the choice of the quality characteristic, the S/N ratio, and control factors and their levels.

- The following guidelines should be used in selecting quality characteristics:

  1. The quality characteristic should be directly related to the energy transfer associated with the basic mechanism or the ideal function of the product.

  2. As far as possible, choose continuous variables as quality characteristics.

  3. The quality characteristics should be monotonic. Also, the related S/N ratio should possess additivity.

  4. Quality characteristics should be easy to measure.

  5. Quality characteristics should be complete—that is, they should cover all dimensions of the ideal function.

  6. For products, having feedback mechanisms, the open loop, sensor and compensation modules should be optimized separately, and the modules should then be integrated. Similarly, complex products should be divided into suitable modules for optimization purposes.

- Although the final success of a product or a process may depend on the reliability or the yield, such responses often do not make good quality characteristics. They tend to cause strong interactions among the control factors as illustrated by the photolithography example.

- Different types of variables can be used as quality characteristics: the output or the response variable, and threshold values of suitable control factors or noise factors for achieving a certain value of the output. When the output is discrete, such as ON-OFF states, it becomes necessary to use the threshold values.

- Additivity of the effects of the control factors is also influenced by the selection of control factors and their levels. If two or more control factors affect the same aspect of the basic phenomenon, then the possibility of interaction among these factors becomes high. When such a situation is recognized, the interaction can be reduced or even eliminated through proper transformation of the control factor levels (sliding levels). A qualitative understanding of how control factors affect a product is important in their selection.

- Selecting a good quality characteristic, S/N ratio, and control factors and their levels is essential in improving the efficiency of development activities. The selection process is not always easy. However, when experiments are conducted using orthogonal arrays, a verification experiment can be used to judge whether the interactions are severe. When interactions are found to be severe, it is possible to look for an improved quality characteristic, S/N ratio, and control factor levels, and, thus, mitigate potential manufacturing problems and field failures.

- Matrix experiment based on an orthogonal array followed by a verification experiment is a powerful tool for detecting lack of additivity. Optimizing a product design one factor at a time does not provide the needed test for additivity.

Chapter 7

CONSTRUCTING
ORTHOGONAL ARRAYS

The benefits of using an orthogonal array to conduct matrix experiments as well as the analysis of data from such experiments are discussed in Chapter 3. The role of orthogonal arrays in a Robust Design experiment cycle is delineated in Chapter 4 with the help of the case study of improving the polysilicon deposition process. This chapter describes techniques for constructing orthogonal arrays that suit a particular case study at hand.

Construction of orthogonal arrays has been investigated by many researchers including Kempthorne [K4], Plackett and Burman [P8], Addelman [A1], Raghavarao [R1], Seiden [S3], and Taguchi [T1]. The process of fitting an orthogonal array to a specific project has been made particularly easy by a graphical tool, called *linear graphs*, developed by Taguchi to represent interactions between pairs of columns in an orthogonal array. This chapter shows the use of linear graphs and a set of standard orthogonal arrays for constructing orthogonal arrays to fit a specific project.

Before constructing an orthogonal array, the following requirements must be defined:

1. Number of factors to be studied

2. Number of levels for each factor

3. Specific 2-factor interactions to be estimated

4. Special difficulties that would be encountered in running the experiments

This chapter describes how to construct an orthogonal array to meet these requirements and consists of the following eleven sections:

- Section 7.1 describes how to determine the minimum number of rows for the matrix experiment by *counting the degrees of freedom.*

- Section 7.2 lists a number of *standard orthogonal arrays* and a procedure for selecting one in a specific case study. A novice to Robust Design may wish to use a standard array that is closest to the needs of the case study, and if necessary, slightly modify the case study to fit a standard array. The remaining sections in this chapter describe various techniques of modifying the standard orthogonal arrays to construct an array to fit the case study.

- Section 7.3 describes the *dummy level method* which is useful for assigning a factor with number of levels less than the number of levels in a column of the chosen orthogonal array.

- Section 7.4 discusses the *compound factor method* which can be used to assign two factors to a single column in the array.

- Section 7.5 describes Taguchi's *linear graphs* and how to use them to assign interactions to columns of the orthogonal array.

- Section 7.6 presents a set of rules for modifying a linear graph to fit the needs of a case study.

- Section 7.7 describes the *column merging method*, which is useful for merging columns in a standard orthogonal array to create columns with larger number of levels.

- Section 7.8 describes *process branching* and shows how to use the linear graphs to construct an appropriate orthogonal array for case studies involving process branching.

- Section 7.9 presents three step-by-step strategies (beginner, intermediate, and advanced) for constructing an orthogonal array.

- Section 7.10 describes the differences between Robust Design and classical statistical experiment design.

- Section 7.11 summarizes the important points of this chapter.

7.1 COUNTING DEGREES OF FREEDOM

The first step in constructing an orthogonal array to fit a specific case study is to count the total degrees of freedom that tells the minimum number of experiments that must be performed to study all the chosen control factors. To begin with, one degree of freedom is associated with the overall mean regardless of the number of control factors to be studied. A 3-level control factor counts for two degrees of freedom because for a 3-level factor, A, we are interested in two comparisons. Taking any one level, A_1, as the base level, we want to know how the response changes when we change the level to A_2 or A_3. In general, *the number of degrees of freedom associated with a factor is equal to one less than the number of levels for that factor.*

The degrees of freedom associated with interaction between two factors, called A and B, are given by the product of the degrees of freedom for each of the two factors. This can be seen as follows. Let n_A and n_B be the number of levels for factors A and B. Then, there are $n_A n_B$ total combinations of the levels of these two factors. From that we subtract one degree of freedom for the overall mean, (n_A-1) for the degrees of freedom of A and (n_B-1) for the degrees of freedom of B. Thus,

Degrees of freedom for interaction A $\times$ B

$$= n_A n_B - 1 - (n_A-1) - (n_B-1)$$

$$= (n_A-1) (n_B-1)$$

$$= \text{(degrees of freedom for A)} \times \text{(degrees of freedom for B)} .$$

Example 1:

Let us illustrate the computation of the degrees of freedom. Suppose a case study has one 2-level factor (A), five 3-level factors (B, C, D, E, F), and we are interested in estimating the interaction A $\times$ B. The degrees of freedom for this experiment are then computed as follows:

Factor/Interaction	Degrees of freedom
Overall mean	1
A	$2-1 = 1$
B, C, D, E, F	$5 \times (3-1) = 10$
A $\times$ B	$(2-1) \times (3-1) = 2$
Total	14

So, we must conduct at least 14 experiments to be able to estimate the effect of each factor and the desired interaction.

7.2 SELECTING A STANDARD ORTHOGONAL ARRAY

Taguchi [T1] has tabulated 18 basic orthogonal arrays that we call *standard orthogonal arrays* (see Appendix C). Most of these arrays can also be found in somewhat different forms in one or more of the following references: Addelman [A1], Box, Hunter, and Hunter [B3], Cochran and Cox [C3], John [J2], Kempthorne [K4], Plackett and Burman [P8], Raghavarao [R1], Seiden [S3], and Diamond [D3]. In many case studies, one of the arrays from Appendix C can be used directly to plan a matrix experiment. An array's name indicates the number of rows and columns it has, and also the number of levels in each of the columns. Thus, the array $L_4(2^3)$ has four rows and three 2-level columns. The array $L_{18}(2^1 3^7)$ has 18 rows; one 2-level column; and seven 3-level columns. Thus, there are eight columns in the array $L_{18}(2^1 3^7)$. For brevity, we generally refer to an array only by the number of rows. When there are two arrays with the same number of rows, we refer to the second array by a prime. Thus, the two arrays with 36 rows are referred to as L_{36} and L'_{36}. The 18 standard orthogonal arrays along with the number of columns at different levels for these arrays are listed in Table 7.1.

TABLE 7.1 STANDARD ORTHOGONAL ARRAYS

Orthogonal Array*	Number of Rows	Maximum Number of Factors	Maximum Number of Columns at These Levels			
			2	3	4	5
L_4	4	3	3	–	–	–
L_8	8	7	7	–	–	–
L_9	9	4	–	4	–	–
L_{12}	12	11	11	–	–	–
L_{16}	16	15	15	–	–	–
L'_{16}	16	5	–	–	5	–
L_{18}	18	8	1	7	–	–
L_{25}	25	6	–	–	–	6
L_{27}	27	13	–	13	–	–
L_{32}	32	31	31	–	–	–
L'_{32}	32	10	1	–	9	–
L_{36}	36	23	11	12	–	–
L'_{36}	36	16	3	13	–	–
L_{50}	50	12	1	–	–	11
L_{54}	54	26	1	25	–	–
L_{64}	64	63	63	–	–	–
L'_{64}	64	21	–	–	21	–
L_{81}	81	40	–	40	–	–

* 2-level arrays: L_4, L_8, L_{12}, L_{16}, L_{32}, L_{64}.
3-level arrays: L_9, L_{27}, L_{81}.
Mixed 2- and 3-level arrays: L_{18}, L_{36}, L'_{36}, L_{54}.

The number of rows of an orthogonal array represents the number of experiments. In order for an array to be a viable choice, the number of rows must be at least equal to the degrees of freedom required for the case study. The number of columns of an array represents the maximum number of factors that can be studied using that array. Further, in order to use a standard orthogonal array directly, we must be able to match the number of levels of the factors with the numbers of levels of the columns in the array. Usually, it is expensive to conduct experiments. Therefore, we use the smallest possible orthogonal array that meets the requirements of the case study. However, in some situations we allow a somewhat larger array so that the additivity of the factor effects can be tested adequately, as discussed in Chapter 8 in conjunction with the differential operational amplifier case study.

Let us consider some examples to illustrate the choice of standard orthogonal arrays.

Example 2:

A case study has seven 2-level factors, and we are only interested in main effects. Here, there are a total of eight degrees of freedom—one for overall mean and seven for the seven 2-level factors. Thus, the smallest array that can be used must have eight or more rows. The array L_8 has seven 2-level columns and, hence, fits this case study perfectly—each column of the array will have one factor assigned to it.

Example 3:

A case study has one 2-level factor and six 3-level factors. This case study has 14 degrees of freedom—one for overall mean, one for the 2-level factor and twelve for the six, 3-level factors. Looking at Table 7.1, we see that the smallest array with at least 14 rows is L_{16}. But this array has fifteen 2-level columns. We cannot directly assign these columns to the 3-level factors. The next larger array is L_{18} which has one 2-level and seven 3-level columns. Here, we can assign the 2-level factor to the 2-level column and the six 3-level factors to six of the seven 3-level columns, keeping one 3-level column empty. *Orthogonality of a matrix experiment is not lost by keeping one or more columns of an array empty.* So, L_{18} is a good choice for this experiment. In a situation like this, we should take another look at the control factors to see if there is an additional control factor to be studied, which we may have ignored as less important. If one exists, it should be assigned to the empty column. Doing this allows us a chance to gain information about this additional factor without spending any more resources.

Example 4:

Suppose a case study has two 2-level and three 3-level factors. The degrees of freedom for this case study are nine. However, L_9 cannot be used directly because it has no 2-level columns. Similarly, the next larger array L_{12} cannot be used directly because it has no 3-level columns. This line of thinking can be extended all the way through the array L_{27}. The smallest array that has at least two 2-level columns and three 3-level columns is L_{36}. However, if we selected L_{36}, we would be effectively wasting $36 - 9 = 27$ degrees of freedom, which would be very inefficient experimentation. This raises the question of whether these standard orthogonal arrays are flexible enough to be modified to accommodate various situations. The answer is yes, and the subsequent sections of this chapter describe the different techniques of modifying orthogonal arrays.

Difficulty in Changing the Levels of a Factor

The columns of the standard orthogonal arrays given in Appendix C are arranged in increasing order of the number of changes; that is, the number of times the level of a factor has to be changed in running the experiments in the numerical order is less for the columns on the left when compared to the columns on the right. Consequently, we should assign a factor whose levels are difficult to change to columns on the left and vice versa.

7.3 DUMMY LEVEL TECHNIQUE

The *dummy level technique* allows us to assign a factor with m levels to a column that has n levels where n is greater than m. Suppose a factor A has two levels, A_1 and A_2. We can assign it to a 3-level column by creating a dummy level A_3 which could be taken the same as A_1 or A_2.

Example 5:

Let us consider a case study that has one 2-level factor (A) and three 3-level factors (B, C, and D) to illustrate the dummy level technique. Here we have eight degrees of freedom. Table 7.2 (a) shows the L_9 array and Table 7.2 (b) shows the experiment layout generated by assigning the factors A, B, C, and D to columns 1, 2, 3, and 4, respectively, and by using the dummy level technique. Here we have taken $A_3 = A_1$ and called it A_1' to emphasize that this is a dummy level.

Note that after we apply the dummy level technique, the resulting array is still *proportionally balanced* and, hence, orthogonal (see Appendix A and Chapter 3). Also, note that in Example 5, we could just as well have taken $A_3 = A_2$. But *to ensure orthogonality, we must consistently take $A_3 = A_1$ or $A_3 = A_2$ within the matrix experiment*. The choice between taking $A_3 = A_1$ or $A_3 = A_2$ depends on many issues. Some of the key issues are as follows:

1. If we take $A_3 = A_2$ then the effect of A_2 will be estimated with two times more precision than the effect of A_1. Thus, the dummy level should be taken to be the one about which we want more precise information. Thus, if A_1 is the starting condition about which we have a fair amount of experience and A_2 is the new alternative, then we should choose $A_3 = A_2$.

TABLE 7.2 DUMMY LEVEL AND COMPOUND FACTOR TECHNIQUES

	(a) L_9 Array					(b) Experiment Layout for Dummy Level Technique (Example 5)					(c) Experiment Layout for Compound Factor Technique (Example 6)			

Expt. No.	Column Number				Expt. No.	Column Number				Expt. No.	Column Number			
	1	2	3	4		1	2	3	4		1	2	3	4
1	1	1	1	1	1	A_1	B_1	C_1	D_1	1	$A_1 E_1$	B_1	C_1	D_1
2	1	2	2	2	2	A_1	B_2	C_2	D_2	2	$A_1 E_1$	B_2	C_2	D_2
3	1	3	3	3	3	A_1	B_3	C_3	D_3	3	$A_1 E_1$	B_3	C_3	D_3
4	2	1	2	3	4	A_2	B_1	C_2	D_3	4	$A_1 E_2$	B_1	C_2	D_3
5	2	2	3	1	5	A_2	B_2	C_3	D_1	5	$A_1 E_2$	B_2	C_3	D_1
6	2	3	1	2	6	A_2	B_3	C_1	D_2	6	$A_1 E_2$	B_3	C_1	D_2
7	3	1	3	2	7	A_1'	B_1	C_3	D_2	7	$A_2 E_1$	B_1	C_3	D_2
8	3	2	1	3	8	A_1'	B_2	C_1	D_3	8	$A_2 E_1$	B_2	C_1	D_3
9	3	3	2	1	9	A_1'	B_3	C_2	D_1	9	$A_2 E_1$	B_3	C_2	D_1
	A	B	C	D		A	B	C	D		AE	B	C	D
	Factor Assignment					Factor Assignment					Factor Assignment			

2. Availability of experimental resources and ease of experimentation also plays a role here. Thus, if A_1 and A_2 are two different raw materials and A_1 is very scarce, then we may choose $A_3 = A_2$ so that the matrix experiment can be finished in a reasonable time.

One can apply the dummy level technique to more than one factor in a given case study. Suppose in Example 5 there were two 2-level factors (A and B) and two 3-level factors (C and D). We can assign the four factors to the columns of the orthogonal array L_9 by taking dummy levels $A_3 = A_1'$ (or $A_3 = A_2'$) and $B_3 = B_1'$ (or $B_3 = B_2'$). Note that the orthogonality is preserved even when the dummy level technique is applied to two or more factors.

The dummy level technique can be further generalized, without losing orthogonality, to assign an m-level factor to an n-level column where m is less than n. For example, for studying the effect of clearance defects and other manufacturing parameters on the reliability of printed wiring boards (described by Phadke, Swann, and Hill [P6], and Mitchell [M1]), a 6-level factor (A) was assigned to a 9-level column by taking $A_7 = A_1'$, $A_8 = A_2'$ and $A_9 = A_3'$.

7.4 COMPOUND FACTOR METHOD

The *compound factor method* allows us to study more factors with an orthogonal array than the number of columns in the array. It can be used to assign two 2-level factors to a 3-level column as follows. Let A and B be two 2-level factors. There are four total combinations of the levels of these factors: A_1B_1, A_2B_1, A_1B_2, and A_2B_2. We pick three of the more important levels and call them as three levels of the compound factor AB. Suppose we choose the three levels as follows: $(AB)_1 = A_1B_1$, $(AB)_2 = A_1B_2$, and $(AB)_3 = A_2B_1$. Factor AB can be assigned to a 3-level column and the effects of A and B can be studied along with the effects of the other factors in the experiment.

For computing the effects of the factors A and B, we can proceed as follows: the difference between the level means for $(AB)_1$ and $(AB)_2$ tells us the effect of changing from B_1 to B_2. Similarly, the difference between the level means for $(AB)_1$ and $(AB)_3$ tells us the effect of changing from A_1 to A_2.

In the compound factor method, however, there is a partial loss of orthogonality. The two compounded factors are not orthogonal to each other. But each of them is orthogonal to every other factor in the experiment. This complicates the computation of the sum of squares for the compounded factors in constructing the ANOVA table. The following examples help illustrate the use of the compound factor method.

Example 6:

Let us go back to Example 4 in Section 7.2 where the case study has two 2-level factors (A and E) and three 3-level factors (B, C, and D). We can form a compound factor AE with three levels $(AE)_1 = A_1E_1$, $(AE)_2 = A_1E_2$ and $(AE)_3 = A_2E_1$. This leads us to four 3-level factors that can be assigned to the L_9 orthogonal array. See Table 7.2(c) for the experiment layout obtained by assigning factors AE, B, C, and D to columns 1, 2, 3, and 4, respectively.

Example 7:

The window photolithography case study described by Phadke, Kackar, Speeney and Grieco [P5] had three 2-level factors (A, B, and D) and six 3-level factors (C, E, F, G, H, and I). The total degrees of freedom for the case study are sixteen. The next larger standard orthogonal array that has several 3-level factors is L_{18} ($2^1 \times 3^7$). The experimenters formed a compound factor BD with three levels $(BD)_1 = B_1D_1$, $(BD)_2 = B_2D_1$ and $(BD)_3 = B_1D_2$. This gave them one 2-level and seven 3-level factors that match perfectly with the columns of the L_{18} array. Reference [P5] also describes the computation of ANOVA for the compound factor method.

As a matter of fact, the experimenters had started the case study with two 2-level factors (A and B) and seven 3-level factors (C through I). However, observing that by dropping one level of one of the 3-level factors, the L_{18} orthogonal array would be suitable, they dropped the least important level of the least important factor, namely factor D. Had they not made this modification to the requirements of the case study, they would have needed to use the L_{27} orthogonal array, which would have amounted to 50 percent more experiments! As illustrated by this example, the experimenter should always consider the possibility of making small modifications in the requirements for saving the experimental effort.

7.5 LINEAR GRAPHS AND INTERACTION ASSIGNMENT

Sections 7.2 through 7.4 considered the situations where we are not interested in estimating any interaction effects. Although in most Robust Design experiments we choose not to estimate any interactions among the control factors, there are situations where we wish to estimate a few selected interactions. The linear graph technique, invented by Taguchi, makes it easy to plan orthogonal array experiments involving interactions.

Confounding of Interactions with Factor Effects

Let us consider the orthogonal array L_8 [Table 7.3 (a)] and suppose we assigned factors A, B, C, D, E, F, and G to the columns 1 through 7, respectively. Suppose we believe that factors A and B are likely to have strong interaction. What effect would the interaction have on the estimates of the effects of the seven factors obtained from this matrix experiment?

The interaction effect is depicted in Figure 7.1. We can measure the magnitude of interaction by the extent of nonparallelism of the effects shown in Figure 7.1. Thus,

$$\text{A} \times \text{B interaction} = (y_{A_2 B_2} - y_{A_1 B_2}) - (y_{A_2 B_1} - y_{A_1 B_1})$$

$$= (y_{A_2 B_2} + y_{A_1 B_1}) - (y_{A_2 B_1} + y_{A_1 B_2}) .$$

From Table 7.3 (a) we see that experiments under level C_1 of factor C (experiments 1, 2, 7 and 8) have combinations $A_1 B_1$ and $A_2 B_2$ of factors A and B; and experiments under level C_2 of factor C (experiments 3, 4, 5 and 6) have combinations $A_1 B_2$ and $A_2 B_1$ of factors A and B. Thus, we will not be able to distinguish the effect of factor C from the A × B interaction. Inability to distinguish effects of factors and interactions is called *confounding*. Here we say that factor C is confounded with interaction A × B. We can avoid the confounding by not assigning any factor to column 3 of the array L_8.

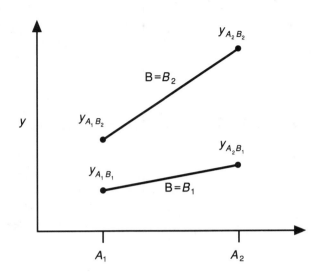

Figure 7.1   2-factor interaction. Interaction between factors A and B shows as nonparallel-ism of the effects of factor A under levels B_1 and B_2 of factor B.

Interaction Table

The *interaction table*, shown in Table 7.3 (b), shows in which column the interaction is confounded with (or contained in) for every pair of columns of the L_8 array. Thus, it can be used to determine which column of the L_8 array should be kept empty (that is, not be assigned to a factor) in order to estimate a particular interaction. From the table, we see that the interaction of columns 1 and 2 is confounded with column 3, the interaction of columns 3 and 5 is confounded with column 6, and so on. Note that the interaction between columns *a* and *b* is the same as that between columns *b* and *a*. That is, the interaction table is a symmetric matrix. Hence, only the upper triangle is given in the table, and the lower triangle is kept blank. Also, the diagonal terms are indicated in parentheses as there is no real meaning to interaction between columns *a* and *a*.

The interaction table contains all the relevant information needed for assigning factors to columns of the orthogonal array so that all main effects and desired interac-tions can be estimated without confounding. The interaction tables for all standard orthogonal arrays prepared by Taguchi [T1] are given in Appendix C, except for the arrays where the interaction tables do not exist, and for the arrays L_{64}, L'_{64} and L_{81}, because they are used rather infrequently. The interaction tables are generated directly from the linear algebraic relations that were used in creating the orthogonal arrays themselves.

TABLE 7.3 L_8 ORTHOGONAL ARRAY AND ITS INTERACTION TABLE

(a) $L_8(2^7)$ orthogonal array

Expt. No.	Column						
	1	**2**	**3**	**4**	**5**	**6**	**7**
1	1	1	1	1	1	1	1
2	1	1	1	2	2	2	2
3	1	2	2	1	1	2	2
4	1	2	2	2	2	1	1
5	2	1	2	1	2	1	2
6	2	1	2	2	1	2	1
7	2	2	1	1	2	2	1
8	2	2	1	2	1	1	2
	A	B	C	D	E	F	G
	Factor Assignment						

(b) Interaction table for L_8

Column	Column						
	1	**2**	**3**	**4**	**5**	**6**	**7**
1	(1)	3	2	5	4	7	6
2		(2)	1	6	7	4	5
3			(3)	7	6	5	4
4				(4)	1	2	3
5					(5)	3	2
6						(6)	1
7							(7)

Note: Entries in this table show the column with which the interaction between every pair of columns is confounded.

Linear Graphs

Using the interaction tables, however, is not very convenient. *Linear graphs* represent the interaction information graphically and make it easy to assign factors and interactions to the various columns of an orthogonal array. In a linear graph, the columns of an orthogonal array are represented by *dots* and *lines*. When two dots are connected by a line, it means that the interaction of the two columns represented by the dots is contained in (or confounded with) the column represented by the line. In a linear graph, each dot and each line has a distinct column number(s) associated with it. Further, every column of the array is represented in its linear graph once and only once.

One standard linear graph for the array L_8 is given in Figure 7.2 (a). It has four dots (or *nodes*) corresponding to columns 1, 2, 4, and 7. Also, it has three lines (or *edges*) representing columns 3, 6, and 5. These lines correspond to the interactions between columns 1 and 2, between columns 2 and 4, and between columns 1 and 4, respectively. From the interaction table, Table 7.3 (b), we can verify that columns 3, 6, and 5 indeed correspond to the interactions mentioned above.

In general, a linear graph does not show the interaction between every pair of columns of the orthogonal array. It is not intended to do so; that information is contained in the interaction table. Thus, the interaction between columns 1 and 3, between columns 2 and 7, etc., are not shown in the linear graph of L_8 in Figure 7.2 (a).

Figure 7.2   **Two standard linear graphs of L_8.**

The other standard linear graph for L_8 is given in Figure 7.2 (b). It, too, has four dots corresponding to columns 1, 2, 4, and 7. Also, it has three lines representing columns 3, 5 and 6. Here, these lines correspond to the interactions between columns 1 and 2, between columns 1 and 4, and between columns 1 and 7, respectively. Let us see some examples of how these linear graphs can be used.

In general, an orthogonal array can have many linear graphs. Each linear graph, however, must be consistent with the interaction table of the orthogonal array. The different linear graphs are useful for planning case studies having different requirements. Taguchi [T1] has prepared many linear graphs, called *standard linear graphs*, for each orthogonal array. Some of the important standard linear graphs are given in Appendix C. Note that the linear graphs for the orthogonal arrays L_{64} and L_{81} are not given in Appendix C because they are needed rather infrequently. However, they can be found in Taguchi [T1]. Section 7.6 describes the rules for modifying linear graphs to fit them to the needs of a given case study.

Example 8:

Suppose in a case study there are four 2-level factors A, B, C, and D. We want to estimate their main effects and also the interactions A × B , B × C, and B × D. Here, the total degrees of freedom are eight, so L_8 is a candidate array. The linear graph in Figure 7.2 (b) can be used directly here. The obvious column assignment is: factor B should be assigned to column 1. Factors A, C, and D can be assigned in an arbitrary order to columns 2, 4, and 7. Suppose we assign factors A, C, and D to columns 2, 4, and 7, respectively. Then the interactions A × B, B × C and B × D can be obtained from columns 3, 5, and 6, respectively. These columns must be kept empty. Table 7.4 shows the corresponding experiment layout.

TABLE 7.4 ASSIGNMENT OF FACTORS AND INTERACTIONS: EXPERIMENT LAYOUT USING ARRAY L_8

Expt. No.	Column*						
	1	2	3	4	5	6	7
1	B_1	A_1		C_1			D_1
2	B_1	A_1		C_2			D_2
3	B_1	A_2		C_1			D_2
4	B_1	A_2		C_2			D_1
5	B_2	A_1		C_1			D_2
6	B_2	A_1		C_2			D_1
7	B_2	A_2		C_1			D_1
8	B_2	A_2		C_2			D_2
	B	A	$A \times B$	C	$B \times C$	$B \times D$	D
	Factor Assignment						

* Note that columns 3, 5, and 6 are left empty (no factors are assigned) so that interactions $A \times B$, $B \times C$ and $B \times D$ can be estimated.

Estimating an interaction means determining the nonparallelism of the factor effects. To estimate an interaction, we prepare a 2-way table from the observed data. For example, to estimate $A \times B$ interaction in Example 8 we prepare the following table whose rows correspond to the levels of factor A, columns correspond to the levels of factor B, and entries correspond to the average response for the particular combination of the levels of factors A and B:

		Level of Factor B	
		B_1	B_2
Level of factor A	A_1	$\dfrac{y_1+y_2}{2}$	$\dfrac{y_5+y_6}{2}$
	A_2	$\dfrac{y_3+y_4}{2}$	$\dfrac{y_7+y_8}{2}$

In the above table y_i stands for the response for experiment i. Experiments 1 and 2 are conducted at levels A_1 and B_1 of factors A and B (see Table 7.4). Accordingly, the entry in the $A_1 B_1$ position in $(y_1 + y_2)/2$. The entries in the other positions of the table are determined similarly. The data of the above 2-way table can be plotted to display the A×B interaction. The interactions B×C and B×D can be estimated in the same manner. If fact, this estimation procedure can be used regardless of the number of levels of a factor.

Example 9:

Suppose there are five 2-level factors A, B, C, D, and E. We want to estimate their main effects and also the interactions A×B and B×C. Here, also, the needed degrees of freedom is eight, making L_8 a candidate array. However, none of the two standard linear graphs of L_8 can be used directly. Section 7.6 shows how the linear graphs can be modified so that a wide variety of experiment designs can be constructed conveniently.

Linear Graphs for 3-level Factors

So far in this section we have discussed the interaction between two 2-level factors. The concept can be extended to situations involving factors with higher number of levels. Figure 7.3 (a) shows an example of no interaction between two 3-level factors, whereas Figures 7.3 (b) and (c) show examples where interaction exists among two 3-level factors.

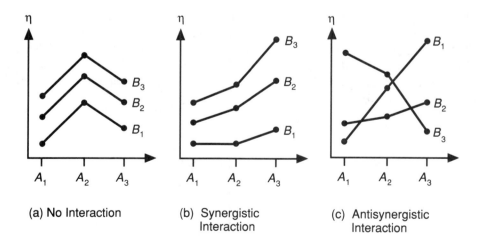

(a) No Interaction   (b) Synergistic Interaction   (c) Antisynergistic Interaction

Figure 7.3   Examples of interaction.

Linear graphs and interaction tables for the arrays L_9, L_{27}, etc., which have 3-level columns, are slightly more complicated than those for arrays with 2-level columns. Each column of a 3-level factor has two degrees of freedom associated with it. The interaction between two 3-level columns has four degrees of freedom. Hence, to estimate the interaction between two 3-level factors, we must keep two 3-level columns empty, in contrast to only one column needed to be kept empty for 2-level orthogonal arrays. This fact is reflected in the interaction tables and linear graphs shown in Appendix C.

As discussed repeatedly in earlier chapters, we generally do not study interactions in Robust Design. Then why study linear graphs? The answer is because linear graphs are useful in modifying orthogonal arrays to fit specific case studies. The following three sections describe the rules for modifying linear graphs and their use in modifying orthogonal arrays.

7.6 MODIFICATION OF LINEAR GRAPHS

The previous section showed how linear graphs can be used to assign main effects and interactions to the columns of standard orthogonal arrays. However, the principal utility of linear graphs is for creating a variety of different orthogonal arrays from the standard ones to fit real problems. The linear graphs are useful for creating 4-level columns in 2-level orthogonal arrays, 9-level columns in 3-level orthogonal arrays and 6-level columns in mixed 2- and 3-level orthogonal arrays. They are also useful for constructing orthogonal arrays for process branching. Sections 7.7 and 7.8 describe these techniques. Common to all these applications of linear graphs is the need to modify a standard linear graph of an orthogonal array so that it matches the linear graph required by a particular problem.

A linear graph for an orthogonal array must be consistent with the interaction table associated with that array; that is, every line in a linear graph must represent the interaction between the two columns represented by the dots it connects. In the following discussion we assume that for 2-level orthogonal arrays, the interaction between columns a and b is contained in column c. Also, the interaction between columns f and g is contained in column c. If it is a 3-level orthogonal array, we assume that the interaction between columns a and b is contained in columns c and d. Also, the interaction between columns f and g is contained in columns c and d. The following three rules can be used for modifying a linear graph to suit the needs of a specific case study.

1. *Breaking a line.* In the case of a 2-level orthogonal array, a line connecting two dots, a and b, can be removed and replaced by a dot. The column associated with this dot is same as the column associated with the line it was created from. In case of linear graphs for 3-level orthogonal arrays, a line has two columns associated with it and it maps into two dots. Figures 7.4 (a) and (b) show this rule diagrammatically.

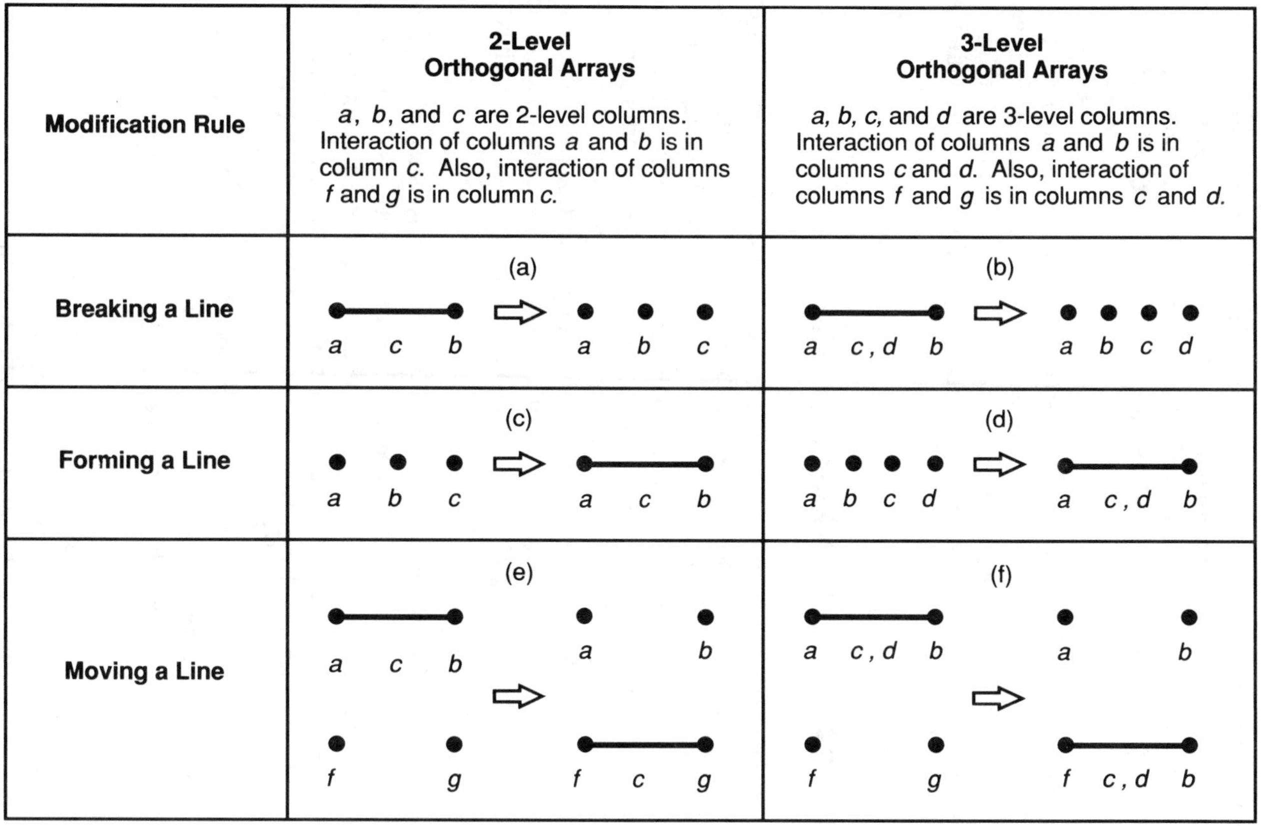
Modification Rule
2-Level Orthogonal Arrays
a, b, and c are 2-level columns. Interaction of columns a and b is in column c. Also, interaction of columns f and g is in column c.
3-Level Orthogonal Arrays
a, b, c, and d are 3-level columns. Interaction of columns a and b is in columns c and d. Also, interaction of columns f and g is in columns c and d.
Breaking a Line
(a)
a c b
a b c
(b)
a c, d b
a b c d
Forming a Line
(c)
a b c
a c b
(d)
a b c d
a c, d b
Moving a Line
(e)
a c b
f g
a b
f c g
(f)
a c, d b
f g
a b
f c, d b

Figure 7.4 Modification of linear graphs.

2. *Forming a line.* A line can be added in the linear graph of a 2-level orthogonal array to connect two dots, *a* and *b*, provided we remove the dot *c* associated with the interaction between *a* and *b*. In the case of the linear graphs for a 3-level orthogonal array, two dots *c* and *d*, which contain the interaction of *a* and *b*, must be removed. The particular dot or dots to be removed can be determined from the interaction table for the orthogonal array. Figures 7.4 (c) and (d) show this rule diagrammatically.

3. *Moving a line.* This rule is really a combination of the preceding two rules. A line connecting two dots *a* and *b* can be removed and replaced by a line joining another set of two dots, say *f* and *g*, provided the interactions $a \times b$ and $f \times g$ are contained in the same column or columns. This rule is diagrammatically shown in Figures 7.4 (e) and (f).

The following examples illustrate the modification of linear graphs.

Example 10:

Consider Example 9 in Section 7.5. The standard linear graph of L_8, Figure 7.5 (a) can be changed into the linear graph shown in Figure 7.5 (b) by breaking the line joining dots 1 and 6. This modified linear graph matches the problem perfectly. The factors A, B, C, D and E should be assigned, respectively, to columns 2, 1, 4, 6, and 7. The $A \times B$ and $B \times C$ interactions can be estimated by keeping columns 3 and 5 empty.

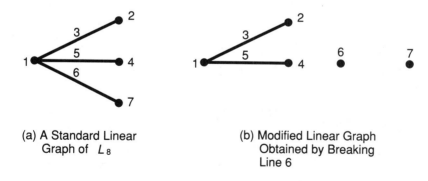

(a) A Standard Linear
Graph of L_8

(b) Modified Linear Graph
Obtained by Breaking
Line 6

Figure 7.5   **Standard and modified linear graph of L_8.**

Example 11:

The purpose of this example is to illustrate the rule 3, namely moving a line. Figure 7.6 (a) shows one of the standard linear graphs of the orthogonal array L_{16}. It can be changed into Figure 7.6 (b) by breaking the line connecting columns 6 and 11, and adding isolated dot for column 13. This can be further turned into Figure 7.6 (c) by adding a line to connect columns 7 and 10, and simultaneously removing the isolated dot 13.

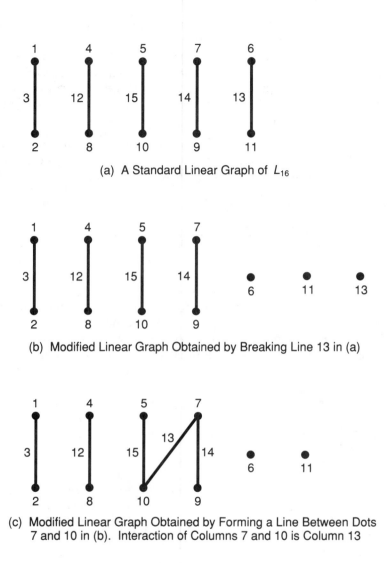

(a) A Standard Linear Graph of L_{16}

(b) Modified Linear Graph Obtained by Breaking Line 13 in (a)

(c) Modified Linear Graph Obtained by Forming a Line Between Dots
   7 and 10 in (b). Interaction of Columns 7 and 10 is Column 13

Figure 7.6   An example of linear graph modification.

7.7 COLUMN MERGING METHOD

The *column merging method* can be used to create a 4-level column in a standard
orthogonal array with all 2-level columns, a 9-level column in a standard orthogonal
array with all 3-level columns, and a 6-level column in a standard orthogonal array
with some 2-level and some 3-level columns.

To create a 4-level column in a standard orthogonal array with 2-level columns, we merge any two columns and their interaction column. For example, in the L_8 array the interaction of columns 1 and 2 lies in column 3. Thus, these three columns can be merged to form a 4-level column. Note that the three columns that are merged have one degree of freedom each, thus together they have the three degrees of freedom needed for the 4-level column.

Suppose columns a, b, and c (the column containing interaction of a and b) are designated to form a 4-level column. The steps in forming the 4-level column are:

1.   Create a new column called abc as follows:

>   For the combination (1,1) in columns a and b write 1 in column abc
>   For the combination (1,2) in columns a and b write 2 in column abc
>   For the combination (2,1) in columns a and b write 3 in column abc
>   For the combination (2,2) in columns a and b write 4 in column abc

2.   Remove columns a, b, and c from the array. These columns cannot be used to study any other factors or interactions.

The creation of a 4-level column using columns 1, 2, and 3 of L_8 is shown in Table 7.5. It can be checked that the resulting array is still balanced and, hence, orthogonal. It can be used to study one 4-level factor and up to four 2-level factors.

TABLE 7.5 COLUMN MERGING METHOD: CREATION OF A 4-LEVEL COLUMN IN L_8.

(a) L_8 array: We designate a = column 1, b = column 2, $a \times b$ interaction is in column 3.

Expt. No.	Column						
	1	2	3	4	5	6	7
1	1	1	1	1	1	1	1
2	1	1	1	2	2	2	2
3	1	2	2	1	1	2	2
4	1	2	2	2	2	1	1
5	2	1	2	1	2	1	2
6	2	1	2	2	1	2	1
7	2	2	1	1	2	2	1
8	2	2	1	2	1	1	2

↑ ↑ ↑
a b $c = a \times b$

(b) Modified L_8 array: Columns 1, 2, and 3 are merged to form a 4-level column.

Expt. No.	Column			
	(1-2-3)	4	5	6	7
1	1	1	1	1	1
2	1	2	2	2	2
3	2	1	1	2	2
4	2	2	2	1	1
5	3	1	2	1	2
6	3	2	1	2	1
7	4	1	2	2	1
8	4	2	1	1	2

In the linear graph of a 2-level orthogonal array we represent a 4-level factor by two dots and a line connecting them.

The column merging procedure above generalizes to orthogonal arrays with columns other than only 2-level columns. Thus, to form a 9-level column in a standard orthogonal array with 3-level columns, we follow the same procedure as above except we must merge four columns: two columns from which we form the 9-level column and the two columns containing their interactions.

7.8 BRANCHING DESIGN

A process for applying covercoat on printed wiring boards consists of (1) spreading the covercoat material (a viscous liquid) on a board, and (2) baking the board to form a hard covercoat layer. Suppose, to optimize this process, we wish to study two types of material (factor A), two methods of spreading (factor B) and two methods of baking (factor C). The two methods of baking are a conventional oven (C_1) and an infrared oven (C_2). For the conventional oven there are two additional control factors, bake temperature (factor D, two levels) and bake time (factor E, two levels). Whereas for the infrared oven, there are two different control factors: infrared light intensity (factor F, two levels) and conveyor belt speed (factor G, two levels).

The factors for the covercoat process are diagrammed in Figure 7.7. Factor C is called a *branching factor* because, depending on its level, we have different control factors for further processing steps. *Branching design* is a method of constructing orthogonal arrays to suit such case studies.

Linear graphs are extremely useful in constructing orthogonal arrays when there is process branching. The linear graph required for the covercoat process is given in Figure 7.8 (a). We need a dot for the branching factor C, and two dots connected with lines to that dot. These two dots correspond to the factors D and E for the conventional oven branch, and F and G for the infrared oven branch. The columns associated with the two interaction lines connected to the branching dot must be kept empty. In the linear graph we also show two isolated dots corresponding to factors A and B.

The standard linear graph for L_8 in Figure 7.8(b) can be modified easily to match the linear graph in Figure 7.8(a). We break the bottom line to form two isolated dots corresponding to columns 6 and 7. Thus, by matching the modified linear graph with the required linear graph, we obtain the column assignment for the control factors as follows:

Factor	Column
A	6
B	7
C	1
D, F	2 (3)
E, G	4 (5)

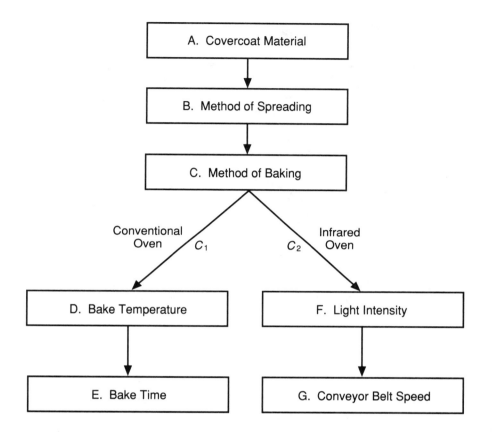

Figure 7.7   Process branching in covercoat process.

Columns 3 and 5, shown in parenthesis, must be kept empty. The factors D and F are assigned to the same column, namely column 2. Whether a particular experiment is conducted by using factor D or F depends on the level of factor C, which is determined by column 1. Thus, the levels of factors D and F are determined jointly by the columns 1 and 2 as follows:

For the combination (1,1) in columns 1 and 2 write D_1 in column 2
For the combination (1,2) in columns 1 and 2 write D_2 in column 2
For the combination (2,1) in columns 1 and 2 write F_1 in column 2
For the combination (2,2) in columns 1 and 2 write F_2 in column 2

Factors D and F can have quite different effects; that is, $m_{D_1} - m_{D_2}$ need not be equal to $m_{F_1} - m_{F_2}$. This difference shows up as interaction between columns 1 and 2, which is contained in column 3. Hence, column 3 must be kept empty. The factors E and G are assigned to the column 4 in a similar way, and column 5 is kept empty.

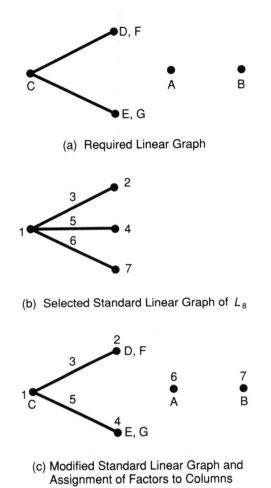

(a) Required Linear Graph

(b) Selected Standard Linear Graph of L_8

(c) Modified Standard Linear Graph and
Assignment of Factors to Columns

Figure 7.8   Covercoat process column assignment through linear graph.

The experiment layout for the covercoat process is given in Table 7.6. Note that experiments 1 through 4 are conducted using the conventional oven, while experiments 5 through 8 are conducted using the infrared oven.

It is possible that after branching, the process can reunite in subsequent steps. Thus, in the printed wiring board application, after the covercoat is applied, we may go through common printing and etching steps that all have a common set of control factors. Branching can also occur in product design; for example, we may select different mechanisms to achieve a part of the function. Here, associated with each mechanism, there would be different control factors.

TABLE 7.6 EXPERIMENT LAYOUT FOR THE COVERCOAT PROCESS

Expt. No.	Column						
	1	**2**	**3**	**4**	**5**	**6**	**7**
1	C_1	D_1		E_1		A_1	B_1
2	C_1	D_1		E_2		A_2	B_2
3	C_1	D_2		E_1		A_2	B_2
4	C_1	D_2		E_2		A_1	B_1
5	C_2	F_1		G_1		A_1	B_2
6	C_2	F_1		G_2		A_2	B_1
7	C_2	F_2		G_1		A_2	B_1
8	C_2	F_2		G_2		A_1	B_2
	C	D,F	Empty	E,G	Empty	A	B
	Factor Assignment						

7.9 STRATEGY FOR CONSTRUCTING AN ORTHOGONAL ARRAY

Up to this point, this chapter discussed many techniques for constructing orthogonal arrays needed by the matrix experiments. This section focuses on showing how to orchestrate the techniques for constructing an orthogonal array to suit a particular case study. The skill needed to apply these techniques varies widely. Accordingly, we describe three strategies—beginner, intermediate, and advanced—requiring progressively higher levels of skill with the techniques described earlier in this chapter. A vast majority of case studies can be taken care of by the beginner and intermediate strategies, whereas a small fraction of the case studies requires the advanced strategy. The router bit life improvement case study in Chapter 11 is one such case study.

Beginner Strategy

A beginner should stick to the direct use of one of the standard orthogonal arrays. Table 7.7 is helpful in selecting a standard orthogonal array to fit a given case study. Because it gets difficult to keep track of data from a larger number of experiments, the beginner is advised to not exceed 18 experiments, which makes the possible choices of orthogonal arrays as L_4, L_8, L_9, L_{12}, L_{16}, L'_{16}, and L_{18}.

TABLE 7.7 BEGINNER STRATEGY FOR SELECTING AN ORTHOGONAL ARRAY

(a) All 2-level Factors

No. of 2-level Factors	Recommended Orthogonal Array
2 – 3	L_4
4 – 7	L_8
8 – 11	L_{12}
12 – 15	L_{16}

(b) All 3-level Factors

No. of 3-level Factors	Recommended Orthogonal Array
2 – 4	L_9
5 – 7	L_{18}*

*When L_{18} is used, one 2-level factor can be used in addition to seven 3-level factors.

A beginner should consider either all 2-level factors or all 3-level factors (preferably 3-level factors) and not attempt to estimate any interactions. This may require him or her to modify slightly the case-study requirements. The rules given in Table 7.7 can then be used to select the orthogonal array.

The assignment of factors to the columns is straightforward in the cases discussed above. Any column can be assigned to any factor, except for factors that are difficult to change, which should be assigned to the columns toward the left.

Among all the arrays discussed above, the array L_{18} is the most commonly used array because it can be used to study up to seven 3-level factors and one 2-level factor, which is the situation with many case studies.

Intermediate Strategy

Experimenters with modest experience in using matrix experiments should use the dummy level, compound factor, and column merging techniques in conjunction with the standard orthogonal arrays to broaden the possible combinations of the factor levels. The factors should have preferably two or three levels and the estimation of interactions should be avoided. Also, as far as possible, arrays larger than L_{18} should be avoided. Table 7.8 can be used to select an appropriate standard orthogonal array depending on the number of 2- and 3-level factors in the case study. The following rules can then be used to modify the chosen standard orthogonal array to fit the case study:

1. To create a 3-level column in the array L_8 or L_{16}, merge three columns in the array (two columns and the column containing their interaction) to form a 4-level column. Then use the dummy level technique to convert the 4-level column into a 3-level column.

2. To create two 3-level columns in the array L_{16}, merge two distinct sets of three columns in the array (two columns and the column containing their interaction) to form two 4-level columns. Then use the dummy level technique to convert the 4-level columns into 3-level columns.

3. When the array L_9 is suggested by the Table 7.8 and the total number of factors is less than or equal to four, use the dummy level technique to assign a 2-level factor to a 3-level column.

4. When the array L_9 is suggested by the Table 7.8 and the total number of factors exceeds four, use the compound factor technique to create a 3-level factor from two 2-level factors until the total number of factors becomes 4.

5. When the array L_{18} is suggested by the Table 7.8 and the number of 2-level columns exceeds one, use the dummy level and compound factor techniques in the manner similar to rules 3 and 4 above.

TABLE 7.8 INTERMEDIATE STRATEGY FOR SELECTING AN ORTHOGONAL ARRAY*

Number of 2-level factors	Number of 3-level factors							
	0	**1**	**2**	**3**	**4**	**5**	**6**	**7**
0			L_9	L_9	L_9	L_{18}	L_{18}	L_{18}
1			L_9	L_9	L_{18}	L_{18}	L_{18}	L_{18}
2	L_4	L_8	L_9	L_9	L_{18}	L_{18}	L_{18}	
3	L_4	L_8	L_9	L_{16}	L_{18}	L_{18}	L_{18}	
4	L_8	L_8	L_9	L_{16}	L_{18}	L_{18}		
5	L_8	L_{16}	L_{16}	L_{16}	L_{18}	L_{18}		
6	L_8	L_{16}	L_{16}	L_{16}	L_{18}			
7	L_8	L_{16}	L_{16}	L_{18}	L_{18}			
8	L_{12}	L_{16}	L_{16}	L_{18}				
9	L_{12}	L_{16}	L_{16}	L_{18}				
10	L_{12}	L_{16}						
11	L_{12}	L_{16}						
12	L_{16}	L_{16}						
13	L_{16}							
14	L_{16}							
15	L_{16}							

* Combination of 2- and 3-level factors not covered by the intermediate strategy are indicated by a blank.

Advanced Strategy

An experimenter, who has a fair amount of experience in conducting matrix experiments and wishes to have wider freedom in terms of the number of factors and their levels or wants to estimate interactions, must use linear graphs and rules for their modification. The advanced strategy consists of the following steps:

1.   Use the beginner or intermediate strategy to obtain a simple solution. If that is not possible, proceed with the following steps.

2.   Count the degrees of freedom to determine the minimum size of the orthogonal array.

3.   Select an appropriate standard orthogonal array from among those listed in Table 7.1. If most of the factors are 2- or 4-level factors, then a 2-level array should be selected. If most of the factors are 3-level factors, then a 3-level array should be selected.

4.   Construct the linear graph required for the case study. The linear graph should contain the interactions to be estimated and also the appropriate patterns for column merging and process branching.

5.   Select a standard linear graph for the chosen array that is closest to the required linear graph.

6.   Modify the standard linear graph to match the required linear graph by using the rules in Section 7.6. The column assignment is obvious when the two linear graphs match. If we do not succeed in matching the linear graphs we must repeat the procedure above with either a different linear graph for the chosen standard orthogonal array, or choose a larger standard orthogonal array, or modify the requirements for the case study.

The advanced strategy needs some skill in using the linear graph modification rules. The router bit life improvement case study of Chapter 11 illustrates the use of the advanced strategy. Artificial intelligence programs can be used to carry out the modifications efficiently as described by Lee, Phadke, and Keny [L1].

7.10 COMPARISON WITH THE CLASSICAL STATISTICAL EXPERIMENT DESIGN

As mentioned in Chapter 1, both classical statistical experiment design and Robust Design use the basic principles of planning experiments and data analysis developed by R. A. Fisher in the 1920s. Thus, there are many common ideas in the two methods. The differences in the methods come about because the two methods were developed by people who were concerned with different problems. This section describes the differences primarily for the benefit of readers familiar with classical statistical experiment design, which is described in many books, such as Box, Hunter, and Hunter [B3], John [J2], Cochran and Cox [C3], Daniel [D1], and Hicks [H2]. It is hoped that this section will help such readers understand and apply the Robust Design Method. This section may be skipped without affecting the readability of the rest of the book.

Any method which was developed over several decades is likely to have variations in the way it is applied. Here, the term classical statistical experiment design refers to the way the method is practiced by the majority of its users. Exceptions to the majority practice are not discussed here. The term Robust Design, of course, means the way it is described in this book.

The comparison is made in three areas: problem formulation, experiment layout, and data analysis. The differences in the areas of experiment layout and data analysis are primarily a result of the fact that the two methods address different problems.

Differences in Problem Formulation

Emphasis on Variance Reduction

The primary problem addressed in classical statistical experiment design is to model the response of a product or process as a function of many factors called *model factors*. Factors, called *nuisance factors*, which are not included in the model, can also influence the response. Various techniques are employed to minimize the effects of the nuisance factors on the estimates of model parameters. These techniques include holding the nuisance factors at constant values during the experiments when possible, as well as techniques called *blocking* and *randomization*. The effects of the nuisance factors not held constant show as variance of the response. Classical statistical experiment design theory is aimed at deriving a mathematical equation relating the mean response to the levels of the model factors. As a general rule, it assumes that the variance of the response remains constant for all levels of the model factors. Thus, it ignores the problem of reducing variability which is critical for quality improvement.

The primary problem addressed in Robust Design is how to reduce the variance of a product's function in the customer's environment. Recall that the variance is caused by the noise factors and the fundamental idea of Robust Design is to find levels of the control factors which minimize sensitivity of the product's function to the noise factors. Consequently, Robust Design is focused on determining the effects of the control factors on the robustness of the product's function. Instead of assuming that the variance of the response remains constant, it capitalizes on the change in variance and looks for opportunities to reduce the variance by changing the levels of the control factors.

In Robust Design, accurate modeling of the mean response is not as important as finding the control factor levels that optimize robustness. This is so because, after the variance has been reduced, the mean response can be easily adjusted with the help of only one control factor. Finding a suitable control factor, known as adjustment factor, which can be used for adjusting the mean response is one of the concerns of the Robust Design method.

Selection of Response/Quality Characteristic

Classical statistical experiment design considers the selection of the response to be outside the scope of its activities, but Robust Design requires a thorough analysis of the engineering scenario in selecting the quality characteristic and the S/N ratio. Guidelines for the selection of the quality characteristic and S/N ratio are given in Chapters 5 and 6.

Frequently, the final goal of a project is to maximize the yield or the percent of products meeting specifications. Accordingly, in classical statistical experiment design yield is often used as a response to be modeled in terms of the model factors. As discussed in Chapters 5 and 6, use of such response variables could lead to unnecessary interactions and it may not lead to a robust product design.

Systematic Sampling of Noise

The two methods also differ in the treatment of noise during problem formulation. Since classical statistical experiment design method is not concerned with minimizing sensitivity to noise factors, the evaluation of the sensitivity is not considered in the method. Instead, noise factors are considered nuisance factors. They are either kept at constant values during the experiments, or techniques called *blocking* and *randomization* are used to block them from having an effect on the estimation of the mathematical model describing the relationship between the response and the model factors.

On the contrary, minimizing sensitivity to noise factors (factors whose levels cannot be controlled during manufacturing or product usage, which are difficult to control, or expensive to control) is a key idea in Robust Design. Therefore, noise factors are systematically sampled for a consistent evaluation of the variance of the quality characteristic and the S/N ratio. Thus, in the polysilicon deposition case study of Chapter 4, the test wafers were placed in specific positions along the length of the reactor and the quality characteristics were measured at specific points on these wafers. This ensures that the effect of noise factors is equitable in all experiments. When there exist many noise factors whose levels can be set in the laboratory, an orthogonal array is used to select a systematic sample, as discussed in Chapter 8, in conjunction with the design of a differential operational amplifier. Use of an orthogonal array for sampling noise is a novel idea introduced by Robust Design and it is absent in classical statistical experiment design.

Transferability of Product Design

Another important consideration in Robust Design is that a design found optimum during laboratory experiments should also be optimum under manufacturing and customer environments. Further, since products are frequently divided into subsystems for design purposes, it is critically important that the robustness of a subsystem be minimally affected by changes in the other subsystems. Therefore, in Robust Design interactions among control factors, especially the antisynergistic interactions, are considered highly undesirable. Every effort is made during problem formulation to select the quality characteristic, S/N ratio, and control factor levels to minimize the interactions. If antisynergistic interactions are discovered during data analysis or verification experiments, the experimenter is advised to go back to Step 1 of the Robust Design cycle and re-examine the choice of quality characteristics, S/N ratios, control factors, noise factors, etc.

On the other hand, classical statistical experiment design has not been concerned with transferability of product design. Therefore, presence of interactions among the model factors is not viewed as highly undesirable, and information gained from even antisynergistic interactions is utilized in finding the factor levels that predict the best average response.

Differences in Experiment Layout

Testing for Additivity

Additivity means absence of all interactions—2-factor, 3-factor, etc. Achieving additivity, though considered desirable, is usually not emphasized in classical statistical experiment design; and orthogonal arrays are not used to test for additivity. Interactions are allowed to be present; they are appropriately included in the model; and experiments are planned so that they can be estimated.

Achieving additivity is very critical in Robust Design, because presence of large interactions is viewed as an indication that the optimum conditions obtained through a matrix experiment may prove to be non-optimum when levels of other control factors (other than those included in the matrix experiment at hand) are changed in subsequent Robust Design experiments. Additivity is considered to be a property that a given quality characteristic and S/N ratio possess or do not possess. Matrix experiment based on an orthogonal array, followed by a verification experiment is used as a tool to test whether the chosen quality characteristic and S/N ratio possess the additivity property.

Efficiency Resulting from Ignoring Interactions among Control Factors

Let us define some terms commonly used in classical statistical experiment design. *Resolution* V designs are matrix experiments where all 2-factor interactions can be estimated along with the main effects. *Resolution* IV designs are matrix experiments where no 2-factor interaction is confounded with the main effects, and no two main effects are confounded with each other. *Resolution* III designs (also called *saturated designs*) are matrix experiments where no two main effects are confounded with each other. In a Resolution III design, 2-factor interactions are confounded with main effects. In an orthogonal array if we allow assigning a factor to each column, then it becomes a Resolution III design. It is possible to construct a Resolution IV design from an orthogonal array by allowing only specific columns to be used for assigning factors.

It is obvious from the above definitions that for a given number of factors, Resolution III design would need the smallest number of experiments, Resolution IV would need somewhat more experiments and Resolution V would need the largest number of experiments. Although heavy emphasis is placed in classical statistical experiment design on ability to estimate 2-factor interactions, Resolution V designs are used only very selectively because of the associated large experimental cost. Resolution IV designs are very popular in classical statistical experiment design. Robust Design almost exclusively uses Resolution III designs, except in some situations where estimation of a few specific 2-factor interactions is allowed.

The relative economics of Resolution III and Resolution IV designs can be understood as follows. By using the interaction tables in Appendix C one can see that Resolution IV designs can be realized in 2-level standard orthogonal arrays by assigning factors to selected columns as shown in Table 7.9.

TABLE 7.9 RESOLUTION III AND IV DESIGNS

Orthogonal Array	Resolution III Design		Resolution IV Design	
	Maximum Number of Factors	Columns to be Used	Maximum Number of Factors	Columns to be Used
L_4	3	1 – 3	2	1,2
L_8	7	1 – 7	4	1,2,4,7
L_{16}	15	1 – 15	8	1,2,4,7,8,11,13,14
L_{32}	31	1 – 31	16	1,2,4,7,8,11,13,14, 16,19,21,22,25,26, 28,31

From the above table it is apparent that for a given orthogonal array roughly twice as many factors can be studied with Resolution III design compared to Resolution IV design.

Screening Experiments

Classical statistical experiment design frequently uses the following strategy for building a mathematical model for the response:

1. *Screening.* Use Resolution III designs to conduct experiments with a large number of model factors for determining whether each of these factors should be included in the mathematical model.

2. *Modeling.* Use Resolution IV (and occasionally, Resolution V) designs to conduct experiments with the factors found important during screening to build the mathematical model.

Robust Design considers screening to be an unnecessary step. Therefore it does not have separate screening and modeling experiments. At the end of every matrix experiment the factor effects are estimated and their optimum levels identified. Robust Design advocates the use of Resolution III designs for all matrix experiments with the exception that sometimes estimation of a few specific 2-factor interactions is allowed.

Flexibility in Experimental Conditions

Because of the heavy emphasis on the ability to estimate interactions and the complexity of the interactions between 3-level factors, classical statistical experiment design is frequently restricted to the use of 2-level fractional factorial designs. Consequently, the number of possible types of experiment conditions is limited. For example, it is not possible to compare three or four different types of materials with a single 2-level fractional factorial experiment. Also, the curvature effect of a factor (see Figure 4.4) cannot be determined with only two levels. However, as discussed earlier in this chapter, the standard orthogonal arrays and the linear graphs used in Robust Design provide excellent flexibility and simplicity in planning multifactor experiments.

Central Composite Designs

Central composite designs are commonly used in classical experiment design, especially in conjunction with the response surface methodology (see Myers [M2]) for estimating the curvature effects of the factors. Although some research is needed to compare the central composite designs with 3-level orthogonal arrays used in Robust Design, the following main differences between them are obvious: the central composite design is useful for only continuous factors, whereas the orthogonal arrays can be used with continuous as well as discrete factors. As discussed in Chapter 3, the predicted response under any combination of the control factor levels has the same variance when an orthogonal array is used. However, this is not true with central composite designs.

Randomization

Running the experiments in a random order is emphasized in classical statistical experiment design to minimize the effect of the nuisance factors on the estimated model factor effects. Running the experiments in an order that minimizes changes in the levels of factors that are difficult to change is considered more important in Robust Design. Randomization is advised to the extent it can be convenient for the experimenter.

In Robust Design, we typically assign control factors to all or most of the columns of an orthogonal array. Consequently, running the experiments in a random order does not scramble the order for all factors effectively. That is, even after arranging the experiments in a random order, it looks as though the experiments are in a nearly systematic order for one or more of the factors.

Nuisance factors are analogous to noise factors in Robust Design terminology. Since robustness against the noise factors is the primary goal of Robust Design, we introduce the noise factors in a systematic sampling manner to permit equitable evaluation of sensitivity to them.

Before we describe the differences in data analysis, we note that many of the layout techniques described in this book can be used beneficially for modeling the mean response also.

Differences in Data Analysis

Two Step Optimization

As mentioned earlier, the differences in data analysis arise from the fact that the two methods were developed to address different problems. One of the common problems in Robust Design is to find control factor settings that minimize variance while attaining the mean on target. In solving this problem, provision must be made to ensure that the solution can be adapted easily in case the target is changed. This is a difficult, multidimensional, constrained optimization problem. The Robust Design method solves it in two steps. First, we maximize the S/N ratio and, then, use a control factor that has no effect on the S/N ratio to adjust the mean function on target. This is an unconstrained optimization problem, much simpler than the original constrained optimization problem. Robust Design addresses many engineering design optimization problems as described in Chapter 5.

Classical statistical experiment design has been traditionally concerned only with modeling the mean response. Some of the recent attempts to solve the engineering design optimization problems in the classical statistical experiment design literature are discussed in Box [B1], Leon, Shoemaker, and Kackar [L2], and Nair and Pregibon [N2].

Significance Tests

In classical statistical experiment design, significance tests, such as the F test, play an important role. They are used to determine if a particular factor should be included in the model. In Robust Design, F ratios are calculated to determine the relative importance of the various control factors in relation to the error variance. Statistical significance tests are not used because a level must be chosen for every control factor regardless of whether that factor is significant or not. Thus, for each factor the best level is chosen depending upon the associated cost and benefit.

7.11 SUMMARY

- The process of fitting an orthogonal array to a specific project has been made particularly easy by the standard orthogonal arrays and the graphical tool, called *linear graphs*, developed by Taguchi to represent interactions between pairs of columns in an orthogonal array. Before constructing an orthogonal array, one must define the requirements which consist of:

    1. Number of factors to be studied

    2. Number of levels for each factor

    3. Specific 2-factor interactions to be estimated

    4. Special difficulties that would be faced in running the experiments

- The first step in constructing an orthogonal array to fit a specific case study is to count the total degrees of freedom that tells the minimum number of experiments that must be performed to study the main effects of all control factors and the chosen interactions.

- Genichi Taguchi has tabulated 18 standard orthogonal arrays. In many problems, one of these arrays can be directly used to plan a matrix experiment. The arrays are presented in Appendix C.

- Orthogonality of a matrix experiment is not lost by keeping empty one or more columns of the array.

- The columns of the standard orthogonal arrays are arranged in the increasing order of number of changes; that is, the number of times the level of a factor must be changed in running the experiments in the numerical order is smaller for the columns on the left than those on the right. Consequently, factors whose levels are difficult to change should be assigned to columns on the left.

- Although in most Robust Design experiments one chooses not to estimate any interactions among the control factors, there are situations where it is desirable to estimate a few selected interactions. The linear graph technique makes it easy to plan orthogonal array experiments that involve interactions.

- Linear graphs represent interaction information graphically and make it easy to assign factors and interactions to the various columns of an orthogonal array. In a linear graph, the columns of an orthogonal array are represented by *dots* and *lines*. When two dots are connected by a line, it means that the interaction of the two columns represented by the dots is contained in (or confounded with) the column(s) represented by the line. In a linear graph, each dot and each line has a distinct column number(s) associated with it. Furthermore, every column of the array is represented in its linear graph once and only once.

- The principal utility of linear graphs is for creating a variety of different orthogonal arrays from the standard orthogonal arrays to fit real problems.

- Techniques described in this chapter for modifying orthogonal arrays are summarized in Table 7.10.

- Depending on the needs of the case study and experience with matrix experiments, the experimenter should use the *beginner*, *intermediate*, or *advanced strategy* to plan experiments. The beginner strategy (see Table 7.7) involves the use of a standard orthogonal array. The intermediate strategy (see Table 7.8) involves minor but simple modifications of the standard orthogonal arrays using the dummy level, compound factor, and column merging techniques. A vast majority of case studies can be handled by the beginner or the intermediate strategies. The advanced strategy requires the use of the linear graph modification rules and is needed relatively infrequently. In complicated case studies, the advanced strategy can greatly simplify the task of constructing orthogonal arrays.

- Although Robust Design draws on many ideas from classical statistical experiment design, the two methods differ because they address different problems. Classical statistical experiment design is used for modeling the mean response, whereas Robust Design is used to minimize the variability of a product's function.

TABLE 7.10 TECHNIQUES FOR MODIFYING ORTHOGONAL ARRAYS

Technique	Application	Needed Linear Graph Pattern
Dummy level	Assign an m-level factor to a n-level column ($m < n$); for example, assign a 2-level factor to a 3-level column.	NA
Compound factor	Assign two 2-level factors to a 3-level column.	NA
Column merging	Create a 4-level column using 2-level columns. Create a 9-level column using 3-level columns. Create a 6-level column using 2- and 3-level columns.	
Branching design	Allow different control factors corresponding to different levels of the branching factor.	

Chapter 8

COMPUTER AIDED ROBUST DESIGN

With the modern advances in engineering sciences, systems of mathematical equations can adequately model the response of many products and processes. Sometimes these equations have a simple closed solution, but in most cases computers must be used to solve them. Nonetheless, costly hardware experiments are not always necessary to design those products or processes.

The Robust Design steps outlined in Chapter 4 for use with hardware experiments can be used just as well to optimize a product or process design when computer models are used to evaluate the response. Of course, some of these steps can be automated with the help of appropriate software, thus making it easier to optimize the design.

This chapter describes the Robust Design of a *differential operational amplifier (op-amp) circuit*, conducted at AT&T Bell Laboratories by Gary Blaine, Joe Leanza, and Madhav Phadke (see Phadke [P3]). The team started with a design that had been optimized by the traditional trial-and-error method and guided by engineering judgment. When the team applied the Robust Design method to the optimized design, an additional reduction in quality loss by a factor of 2.7 occurred with no increase in manufacturing cost.

The differential op-amp example helps illustrate optimization of a product design, particularly when a computer model is available. It also shows how to use an orthogonal array to simulate the variation in noise factors and, thus, estimate the quality loss

and manufacturing yield. Also, it brings out some important considerations in tolerance design and helps illustrate the use of an orthogonal array to select an appropriate
signal-to-noise (S/N) ratio for a project.

This chapter has eleven sections.

- Section 8.1 describes the differential op-amp circuit and its main function (Step 1
  of Robust Design steps described in Chapter 4).

- Section 8.2 discusses the noise factors and their statistical properties (Step 2).

- Section 8.3 summarizes some commonly used methods of simulating the effect
  of variation in noise factors.

- Section 8.4 discusses the orthogonal array used in determining the testing conditions and the evaluation of the effect of noise factors for our circuit (Step 2).

- Section 8.5 gives the S/N ratio used in this example (Step 3).

- Section 8.6 describes the control factors, their alternate levels, and the use of an
  orthogonal array for finding optimum settings of control factors (Steps 4 through
  7). It also describes the verification experiment and the final results (Step 8).

- Section 8.7 discusses tolerance design.

- Section 8.8 describes ways of reducing simulation effort.

- Section 8.9 analyzes the nonlinearity of the circuit.

- Section 8.10 describes the role of orthogonal arrays in selecting an appropriate
  S/N ratio.

- Section 8.11 summarizes the important points of this chapter.

8.1 DIFFERENTIAL OP-AMP CIRCUIT

Differential op-amp circuits are commonly used in telecommunications. The particular
application the design team was interested in was to use the circuit as a preamplifier in
coin telephones. An important quality characteristic for this circuit is its offset voltage.
The ideal value for offset voltage is zero volts. If the offset voltage is large, regardless
of whether it is negative or positive, then the circuit cannot be used for coin telephones
located far from a central office. Therefore, it is important to keep the offset voltage
close to zero under all noise conditions.

The circuit diagram for the differential op-amp circuit is given in Figure 8.1.
There are two current sources (OCS, CPCS), five transistors, and eight resistors. The

circuit has a balancing property (symmetry) that requires the following relationships among the nominal values of the various circuit parameters:

$$RFP = RFM$$
$$RPEP = RPEM$$
$$RNEP = RNEM$$
$$AFPP = AFPM$$
$$AFNP = AFNM$$
$$SIEPP = SIEPM$$
$$SIENP = SIENM$$

$$(8.1)$$

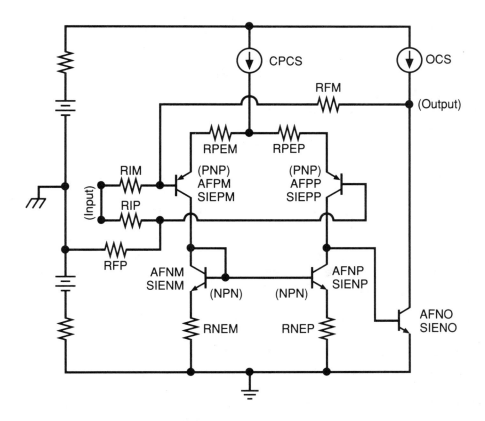

Figure 8.1   Differential operational amplifier circuit. This input is shorted for evaluating the dc offset voltage.

The circuit parameter names beginning with AF refer to the alpha parameters of the transistors, and those beginning with SIE are the saturation currents for the transistors. Further, the gain requirement of the circuit dictates the following ratios of resistance values:

$$RIM = RFM/3.55 \qquad\qquad (8.2)$$
$$RIP = RFM/3.55$$

These relationships among the circuit parameters, expressed by Equations (8.1) and (8.2), are called *tracking relationships*. Many product architectures include a variety of tracking relationships among the product parameters. Past experiences with similar products are often specified through such relationships.

8.2 DESCRIPTION OF NOISE FACTORS

Because of the symmetric architecture of the circuit, the dc offset voltage is nearly zero if all circuit parameters are exactly at nominal values; however, that is not the case in practice. Manufacturing variations violate the symmetry, which leads to high offset voltage. So, the deviations in circuit parameter values from the respective nominal values are the primary noise factors for this circuit. Another important noise factor is the ambient temperature because the circuit is expected to function outdoors in many climatically different areas. Thus, there are 21 noise parameters for this circuit: deviations in the 20 circuit parameters from their nominal values and the temperature variation.

The nature of the variation of product parameters is greatly influenced by the manufacturing process. This statement is also true for the differential op-amp circuit. Two important aspects of variation we need to consider are:

1. *Shape of the distribution.* For some circuit parameters, the manufacturing variation is roughly symmetric around the nominal and it resembles the normal distribution, whereas for other parameters there is a long-tailed distribution resembling a log-normal distribution. See Figure 8.2(a)−(b).

2. *Correlation.* When the circuits are made as an integrated circuit, which is the case with this differential op-amp circuit, values of certain components can have a high correlation. See Figure 8.2(c).

For the starting design in the case study, the mean values and the tolerances on the noise factors are listed in Table 8.1. The listed tolerances are the three standard-deviation limits. Thus, for RPEM, the standard deviation is $21/3 = 7$ percent of its nominal value. All saturation currents in these transistors have a long-tailed distribution. Accordingly, the tolerances on these parameters are specified as a multiple. Thus, we approximate the distribution of SIEPM as follows: $\log_{10}$ (SIEPM) has mean value $\log_{10}(3 \times 10^{-13})$ and standard deviation $(\log_{10} 7)/3$.

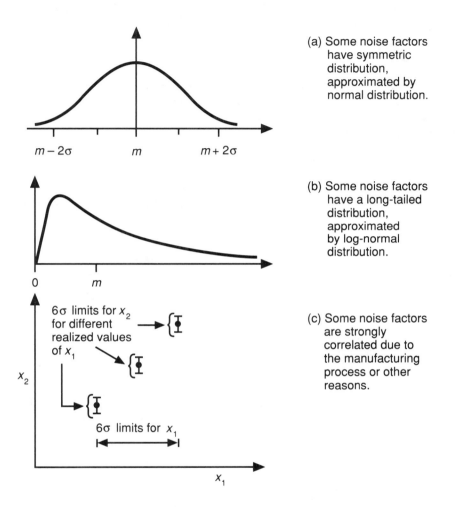

(a) Some noise factors have symmetric distribution, approximated by normal distribution.

(b) Some noise factors have a long-tailed distribution, approximated by log-normal distribution.

(c) Some noise factors are strongly correlated due to the manufacturing process or other reasons.

Figure 8.2   Distributions of noise factors.

In this circuit, the mean values of only the parameters 1 through 5 can be independently specified by the designer. The other mean values are determined by either the tracking relationships or the chosen manufacturing technology. Consider the resistance RPEP. Its nominal value is equal to the nominal value of RPEM. Further, these resistors are located close to each other on a single chip. As a result, there is less variation of RPEP around RPEM on the same chip when compared to the variation of either of these resistances from chip to chip on the same wafers or across wafers. This correlation is expressed by a large tolerance (21 percent) on RPEM and a small tolerance (2 percent) on RPEP around the value of RPEM. Figure 8.2(c) shows the correlation in a graphical form. Suppose in a particular group of circuits RPEM = 15 kΩ.

Then, for that group of circuits RPEP will vary around 15 kΩ with three standard-deviation limits of 2 percent of 15 kΩ. If for another group of circuits RPEM is 16.5 kΩ (10 percent over 15 Ω), then RPEP will vary around 16.5 kΩ with three standard-deviation limits equal to 2 percent of 16.5 kΩ.

The correlations among the other circuit parameters are specified in a similar manner in Table 8.1. RFP, RIM, and RIP are correlated with RFM; RPEP with RPEM; RNEP with RNEM; AFPP with AFPM; AFNP with AFNM; SIEPP with SIEPM; and SIENP with SIENM.

TABLE 8.1 NOISE FACTORS FOR THE DIFFERENTIAL OP-AMP CIRCUIT

Noise Factor	Mean	Tolerance†	Levels (Multiply by Mean)		
			1	**2**	**3**
1. RFM*	71 kΩ	1%	0.9967	1.0033	
2. RPEM*	15 kΩ	21%	0.93	1.07	
3. RNEM*	2.5 kΩ	21%	0.93	1.07	
4. CPCS*	20 μA	6%	0.98	1.02	
5. OCS*	20 μA	6%	0.98	1.02	
6. RFP	RFM	2%	0.9933	1.0067	
7. RIM	RFM/3.55	2%	0.9933	1.0067	
8. RIP	RFM/3.55	2%	0.9933	1.0067	
9. RPEP	RPEM	2%	0.9933	1.0067	
10. RNEP	RNEM	2%	0.9933	1.0067	
11. AFPM	0.9817	2.5%	0.99	1	1.01
12. AFPP	AFPM	0.5%	0.998	1	1.002
13. AFNM	0.971	2.5%	0.99	1	1.01
14. AFNP	AFNM	0.5%	0.998	1	1.002
15. AFNO	0.975	1%	0.99	1	1.01
16. SIEPM	3.0E−13A	Factor of 7	0.45	1	2.21
17. SIEPP	SIEPM	Factor of 1.214	0.92	1	1.08
18. SIENM	6.0E−13A	Factor of 7	0.45	1	2.21
19. SIENP	SIENM	Factor of 1.214	0.92	1	1.08
20. SIENO	6.0E−13A	Factor of 2.64	0.67	1	1.49
21. TKELV	298°K	15%	0.94	1	1.06

* The mean values for these parameters can be set by the designer. The values shown in this table refer to a particular design, called starting design.

† Tolerance means three-standard-deviation limit.

8.3 METHODS OF SIMULATING THE VARIATION IN NOISE FACTORS

Let us denote the noise factors by $x_1, \cdots, x_k$ where $k = 21$. Suppose all these noise factors meet the mean and tolerances, including the implied correlations, specified by Table 8.1. How can we evaluate the mean and variance of the offset voltage, which is the product's response being considered? Three common methods of evaluating the mean and variance of a product's response resulting from variations in many noise factors are Monte Carlo simulation, Taylor series expansion, and orthogonal array based simulation. They are described briefly next.

Monte Carlo Simulation

In this method, a random number generator is used to simulate a large number of combinations of the noise factors, called *testing conditions*. The value of the response is computed for each testing condition, and the mean and variance of the response are then calculated. For obtaining accurate estimates of mean and variance, the Monte Carlo method requires evaluation of the response under a large number of testing conditions. This can be very expensive, especially if we also want to compare many combinations of control factor levels.

Taylor Series Expansion

In this method, the mean response is estimated by setting each noise factor equal to its nominal value. To estimate the variance of the response, we find the derivatives of the response with respect to each noise factor. Let v denote the offset voltage and $\sigma_1^2, \cdots, \sigma_k^2$ denote the variances of the k noise factors. The variance of v is then computed by the following formula:

$$\sigma_v^2 = \sum_{i=1}^{k} \left(\frac{\partial v}{\partial x_i} \right)^2 \sigma_i^2 . \tag{8.3}$$

Note that the derivatives used in this formula can be evaluated mathematically or numerically. Equation (8.3), which is based on first-order Taylor series expansion, gives quite accurate estimates of variance when the correlations among the noise factors are negligible and the tolerances are small, so that interactions among the noise factors and the higher order terms are negligible. Otherwise, higher order Taylor series expansions must be used, which makes the formula for evaluating the variance of the response quite complicated and computationally expensive. Thus, Equation (8.3) does not always give an accurate estimate of variance.

Orthogonal Array Based Simulation

In this method, proposed by Taguchi, orthogonal arrays are used to sample the domain of noise factors. For each noise variable, we take either two or three levels. Suppose μ_i and σ_i^2 are the mean and variance, respectively, for the noise variable x_i. When we take two levels, we choose them to be $\mu_i - \sigma_i$ and $\mu_i + \sigma_i$. Note that the mean and variance of these two levels are μ_i and σ_i^2, respectively. Similarly, when we take three levels, we choose them to be $\mu_i - \sqrt{3/2}\ \sigma_i$, μ_i, and $\mu_i + \sqrt{3/2}\ \sigma_i$. Here also, the mean and variance of the three levels are μ_i and σ_i^2, respectively. We then assign the noise factors to the columns of an orthogonal array to determine testing conditions (sampling points) for evaluating the response. From these values of response, we can estimate the mean and variance of the response. Note that orthogonal array based simulation can be performed with hardware experiments as well, provided experimental equipment allows us to set the levels of noise factors.

The advantage of this method over the Monte Carlo method is that it needs a much smaller (orders of magnitude smaller) number of testing conditions; yet, the accuracy is excellent. Next, the orthogonal array based simulation gives common testing conditions for comparing two or more combinations of control factor settings. If different seeds are used in generating the random numbers, we do not get common testing conditions. Further, when interactions and correlations among the noise factors are strong, an orthogonal array based simulation gives more accurate estimates of mean and variance compared to Taylor series expansion. On the other hand, when the interactions and correlations are small, both the Taylor series expansion method and the orthogonal array based simulation method give the same results.

8.4 ORTHOGONAL ARRAY BASED SIMULATION OF VARIATION IN NOISE FACTORS

Selecting three levels rather than two levels for the noise factors gives more accurate estimates of variance. However, selecting two levels leads to a smaller orthogonal array and, hence, a saving in the simulation cost. In the differential op-amp application, the design team wanted to limit the size of the orthogonal array for simulating the variation in noise factors to L_{36}. Note that the L_{36} array, shown in Table 8.2, has eleven 2-level columns and twelve 3-level columns. So, two levels were taken for the first ten factors (the eight resistors and the two current sources), and three levels were taken for the remaining eleven factors (the ten transistor parameters and the temperature).

The levels for the various noise parameters are shown in Table 8.1. The levels shown are the nominal value plus the deviation from the nominal value. Thus, for the parameter RPEM, $\sigma = 21/3 = 7$ percent. So, level 1 is $\mu - 0.07\mu = 0.93\mu$, and level 2 is $\mu + 0.07\mu = 1.07\mu$, where μ is the mean value for RPEM. Due to the correlation induced by the manufacturing process, the mean value of RPEP is equal to the realized value of RPEM. The σ for the variation of RPEP around RPEM is $2/3 = 0.67$ percent. Thus, the two levels of RPEP are 0.9933 and 1.0067 times the realized value of RPEM. *Thus, we take care of the correlation through sliding levels.*

TABLE 8.2 NOISE ORTHOGONAL ARRAY*

Expt. No.	1	2	3	4	5	6	7	8	9	10	11	12	13	14	15	16	17	18	19	20	21	22	23	Offset Voltage‡ (mV)
1	1	1	1	1	1	1	1	1	1	1	1	1	1	1	1	1	1	1	1	1	1	1	1	−22.8
2	1	1	1	1	1	1	1	1	1	1	1	2	2	2	2	2	2	2	2	2	2	2	2	−4.8
3	1	1	1	1	1	1	1	1	1	1	1	3	3	3	3	3	3	3	3	3	3	3	3	14.2
4	1	1	1	1	1	2	2	2	2	2	2	1	1	1	1	2	2	2	2	3	3	3	3	16.0
5	1	1	1	1	1	2	2	2	2	2	2	2	2	2	2	3	3	3	3	1	1	1	1	−55.8
6	1	1	1	1	1	2	2	2	2	2	2	3	3	3	3	1	1	1	1	2	2	2	2	37.7
7	1	1	2	2	2	1	1	1	2	2	2	1	1	2	3	1	2	3	3	1	2	2	3	−16.5
8	1	1	2	2	2	1	1	1	2	2	2	2	2	3	1	2	3	1	1	2	3	3	1	20.8
9	1	1	2	2	2	1	1	1	2	2	2	3	3	1	2	3	1	2	2	3	1	1	2	−9.1
10	1	2	1	2	2	1	2	2	1	1	2	1	1	3	2	1	3	2	3	2	1	3	2	33.5
11	1	2	1	2	2	1	2	2	1	1	2	2	2	1	3	2	1	3	1	3	2	1	3	4.9
12	1	2	1	2	2	1	2	2	1	1	2	3	3	2	1	3	2	1	2	1	3	2	1	−56.7
13	1	2	2	1	2	2	1	2	1	2	1	1	2	3	1	3	2	1	3	3	2	1	2	25.2
14	1	2	2	1	2	2	1	2	1	2	1	2	3	1	2	1	3	2	1	1	3	2	3	−40.6
15	1	2	2	1	2	2	1	2	1	2	1	3	1	2	3	2	1	3	2	2	1	3	1	−12.4
16	1	2	2	2	1	2	2	1	2	1	1	1	2	3	2	1	1	3	2	3	3	2	1	61.3
17	1	2	2	2	1	2	2	1	2	1	1	2	3	1	3	2	2	1	3	1	1	3	2	−38.5
18	1	2	2	2	1	2	2	1	2	1	1	3	1	2	1	3	3	2	1	2	2	1	3	−15.5
19	2	1	2	2	1	1	2	2	1	2	1	1	2	1	3	3	3	1	2	2	1	2	3	−29.2
20	2	1	2	2	1	1	2	2	1	2	1	2	3	2	1	1	1	2	3	3	2	3	1	27.2
21	2	1	2	2	1	1	2	2	1	2	1	3	1	3	2	2	2	3	1	1	3	1	2	−31.4
22	2	1	2	1	2	2	2	1	1	1	2	1	2	2	3	3	1	2	1	1	3	3	2	−48.1
23	2	1	2	1	2	2	2	1	1	1	2	2	3	3	1	1	2	3	2	2	1	1	3	13.4
24	2	1	2	1	2	2	2	1	1	1	2	3	1	1	2	2	3	1	3	3	2	2	1	21.8
25	2	1	1	2	2	2	1	2	2	1	1	1	3	2	1	2	3	3	1	3	1	2	2	19.7
26	2	1	1	2	2	2	1	2	2	1	1	2	1	3	2	3	1	1	2	1	2	3	3	−19.0
27	2	1	1	2	2	2	1	2	2	1	1	3	2	1	3	1	2	2	3	2	3	1	1	8.2
28	2	2	2	1	1	1	1	2	2	1	2	1	3	2	2	2	1	1	3	2	3	1	3	10.8
29	2	2	2	1	1	1	1	2	2	1	2	2	1	3	3	3	2	2	1	3	1	2	1	40.1
30	2	2	2	1	1	1	1	2	2	1	2	3	2	1	1	1	3	3	2	1	2	3	2	−40.1
31	2	2	1	2	1	2	1	1	1	2	2	1	3	3	3	2	3	2	2	1	2	1	1	−31.1
32	2	2	1	2	1	2	1	1	1	2	2	2	1	1	1	3	1	3	3	2	3	2	2	−56.6
33	2	2	1	2	1	2	1	1	1	2	2	3	2	2	2	1	2	1	1	3	1	3	3	48.9
34	2	2	1	1	2	1	2	1	2	2	1	1	3	1	2	3	2	3	1	2	2	3	1	−46.5
35	2	2	1	1	2	1	2	1	2	2	1	2	1	2	3	1	3	1	2	3	3	1	2	66.2
36	2	2	1	1	2	1	2	1	2	2	1	3	2	3	1	2	1	2	3	1	1	2	3	−22.7
	1	2	3	4	5	6	7	8	9	10	e	11	12	13	14	15	16	17	18	19	20	21	e	

Noise Factor Assignment†

* Columns 1–23 form the L_{36} orthogonal array. Columns 1–10 and 12–22 form the noise orthogonal array.

† Empty column is identified by e.

‡ The values in this column correspond to the starting design.

Let us now see how the levels for the transistor parameter SIEPM are calculated. As mentioned earlier, this parameter has a long-tailed distribution that could be approximated by the log-normal density. Let $\log_{10}(\mu_{16})$ be the mean for $\log_{10}(\text{SIEPM})$. Then, $\log_{10}(\text{SIEPM}/\mu_{16})$ has mean zero and standard deviation equal to $(\log_{10}7)/3 = 0.2817$. So, level 1 for $\log_{10}(\text{SIEPM}/\mu_{16}) = -\sqrt{3/2}\,(0.2817) = -0.345$. Thus, in the natural scale, the level 1 for SIEPM is $10^{-0.345} = 0.45$ times μ_{16}. Similarly, the level 3 for SIEPM is $10^{0.345} = 2.21$ times μ_{16}, while level 2 is equal to μ_{16}. The levels for the other noise parameters are calculated in a similar manner.

The factors 1 through 10 were assigned to columns 1 through 10 of the L_{36} array, respectively, and the factors 11 through 21 were assigned to columns 12 through 22, respectively. Columns 11 and 23 were kept empty. The submatrix of L_{36} formed by columns 1 through 10 and 12 through 21 is called the *noise orthogonal array*, as it is used to simulate the effect of noise factors.

The 36 rows of the noise orthogonal array represent 36 testing conditions. We refer to Table 8.1 to determine the settings of the circuit parameters corresponding to an individual row of the noise orthogonal array. Consider the starting design. As indicated in Table 8.1, it is characterized by the following circuit parameter values: RFM = 71 kΩ, RPEM = 15 kΩ, RNEM = 2.5 kΩ, CPCS = 20μA, and OCS = 20 μA. For the starting design, the computer model for circuit analysis was used to calculate the offset voltage v for each row of the noise orthogonal array. These values are listed in Table 8.2 along with the noise orthogonal array. The average of these 36 values is -3.54 mV and the variance is 1184 (mV)^2. The corresponding standard deviation is 34.4 mV.

It is obvious that the quality loss, denoted by Q, for this circuit design is given by

$$
\begin{aligned}
Q &= k \times (\text{mean square offset voltage}) \\
&= k \times (3.54^2 + 34.4^2) \\
&= 1197\ k.
\end{aligned}
$$

where k is the quality loss coefficient. Suppose the maximum allowed variation for the offset voltage is 0 ± 35 mV. Then, assuming that the distribution of the offset voltage is approximately normal (gaussian) with mean and variance as determined earlier, the percent yield (p) of circuits produced under this design would be given by

$$
p = \Phi\left[\frac{35 + 3.54}{34.4}\right] - \Phi\left[\frac{-35 + 3.54}{34.4}\right]
$$

$$
= \Phi(1.12) - \Phi(-0.91)
$$

$$
= 0.87 - 0.18 = 0.69 = 69 \text{ percent.}
$$

In this formula, Φ is the cumulative distribution function of the normal (gaussian) distribution. Thus, by using a small number of evaluations of the offset voltage, we are able to estimate the quality loss and the manufacturing yield.

The offset voltages evaluated under the 36 testing conditions can be further analyzed by the ANOVA technique discussed in Chapter 3 to determine the contribution of each noise factor. This is an added benefit of orthogonal array based noise simulation over the Monte Carlo simulation method. The results of ANOVA are shown in Table 8.3. We can see that the four largest contributors to the variance are SIENP, AFNO, AFNM, and SIEPP. Together, they account for 95.5 percent of the variance of the offset voltage. Thus, the remaining noise factors have a negligible contribution to the variance of the offset voltage.

TABLE 8.3 ANOVA FOR THE OFFSET VOLTAGE UNDER THE STARTING DESIGN*

Noise Factor	Level Means (mV)			Degrees of Freedom	Sum of Squares
	1	**2**	**3**		
1. RFM	− 3.288	− 3.796		1	2
2. RPEM	− 3.208	− 3.876		1	4
3. RNEM	− 3.705	− 3.379		1	1
4. CPCS	− 2.689	− 4.395		1	26
5. OCS	− 3.872	− 3.212		1	4
6. RFP	− 3.452	− 3.632		1	0
7. RIM	− 3.620	− 3.464		1	0
8. RIP	− 3.648	− 3.436		1	0
9. RPEP	− 8.027	0.943		1	724
10. RNEP	− 0.984	− 6.100		1	236
11. AFPM	− 2.315	− 3.558	− 4.754	2	36
12. AFPP	0.288	− 2.621	− 8.293	2	457
13. AFNM	− 19.364	− 3.088	11.827	2	5841
14. AFNP	− 7.674	− 2.578	− 0.374	2	336
15. AFNO	14.701	− 3.908	− 21.419	2	7831
16. SIEPM	− 4.060	− 3.562	− 3.004	2	7
17. SIEPP	5.420	− 3.906	− 12.141	2	1853
18. SIENM	− 2.731	− 3.800	− 4.096	2	12
19. SIENP	− 35.278	− 3.369	28.021	2	24041
20. SIENO	− 2.894	− 4.737	− 2.995	2	26
21. TKELV	− 3.068	− 3.878	− 3.680	2	4
Error				3	8
Total				35	41449

* Overall mean = − 3.54 mV

8.5 QUALITY CHARACTERISTIC AND S/N RATIO

The selection of offset voltage as the quality characteristic was quite straightforward in this project and was a natural choice, as it intuitively satisfies the guidelines given in Chapter 6. The ideal value for the offset voltage is 0.0 mV. Depending on the particular values of the circuit parameters, the offset voltage can be either positive or negative. The design of the differential op-amp circuit for offset voltage is clearly a static problem. Referring to the classification of static problems given in Chapter 5, the current problem can be classified as a signed-target type of problem. Thus, the appropriate S/N ratio, η, to be maximized is

$$\eta = -10 \log_{10} \text{ (variance of the offset voltage).}$$

If under the control factor settings that maximize η we get a nonzero mean offset voltage, we can take care of it in one of two ways: (1) subtract the mean voltage in the circuit that receives the output of the differential op-amp circuit, or (2) find a control factor of the differential op-amp circuit that has a negligible or no effect on η, but has an appreciable effect on the mean, and use it to adjust the mean at zero.

Notice that there is always a potential to misclassify a nominal-the-best type problem as a signed-target problem by subtracting the nominal value. Here, we are not making such a mistake because zero offset voltage is a naturally occurring value. One way to test whether a problem is being misclassified as signed-target is to see whether the variance is identically zero for a particular value of the quality characteristic. If it is, then by shifting the origin to that value, the problem should be classified as a nominal-the-best type. Note that it is also necessary to ensure that after shifting the origin, the quality characteristic takes only positive values. Recall that we followed this strategy for the heat exchanger example in Chapter 6.

It should also be observed that we should not classify this problem as a smaller-the-better type because the offset voltage can be positive as well as negative. The selection of an appropriate S/N ratio is discussed further in Section 8.10.

8.6 OPTIMIZATION OF THE DESIGN

Control Factors and Their Levels

As mentioned earlier, the circuit designer can set values of only five parameters: RFM, RPEM, RNEM, CPCS, and OCS. These parameters constitute the control factors. The other parameters are determined by the tracking relationships and the manufacturing process. When the design team started the project, the circuit was "optimized" intuitively to get low offset voltage. The corresponding values of the control factors were

RFM = 71 kΩ, RPEM = 15 kΩ, RNEM = 2.5 kΩ, CPCS = 20 μA, and OCS = 20 μA. For each of these control factors, three levels were taken as shown in Table 8.4. For each control factor, level 2 is the starting level, level 1 is one-half of the starting level, and level 3 is two times the starting level. Thus, we include a wide range of values with these levels. This is necessary to benefit adequately from the nonlinearity of the relationship between the control factors, noise factors, and the offset voltage.

TABLE 8.4 CONTROL FACTORS FOR THE DIFFERENTIAL OP-AMP CIRCUIT

			Levels*		
Name		**Description**	**1**	**2**	**3**
A.	RFM	Feedback resistance, minus terminal (kΩ)	35.5	71	142
B.	RPEM	Emitter resistance, PNP minus terminal (kΩ)	7.5	15	30
C.	RNEM	Emitter resistance, NPN minus terminal (kΩ)	1.25	2.5	5
D.	CPCS	Complementary pair current source (μA)	10	20	40
E.	OCS	Output current source (μA)	10	20	40

* The starting levels are indicated by an underscore.

Control Orthogonal Array

From the discussion in Chapter 7, it is obvious that the orthogonal array L_{18} is the best choice for studying five control factors at three levels each. This is because there are $5 \times 2 + 1 = 11$ degrees of freedom associated with the five control factors, and L_{18} is the smallest orthogonal array with at least five 3-level columns. However, at the time the study was conducted, the L_{36} array was used, because computer time was not an issue. The factors RFM, RPEM, RNEM, CPCS, and OCS were assigned to columns 12, 13, 14, 15, and 16, respectively. The submatrix of L_{36}, formed by columns 12 through 16, is called the *control orthogonal array* because it is used to determine trial combinations of control factor settings. The control orthogonal array is shown in Table 8.5. Note that both the noise orthogonal array and the control orthogonal array are derived from a single standard orthogonal array L_{36}. This need not be the situation with all computer aided design applications.

TABLE 8.5 CONTROL ORTHOGONAL ARRAY*

Expt. No.	Column No. and Factor Assignment					Expt. No.	Column No. and Factor Assignment				
	1 A	2 B	3 C	4 D	5 E		1 A	2 B	3 C	4 D	5 E
1	1	1	1	1	1	19	1	2	1	3	3
2	2	2	2	2	2	20	2	3	2	1	1
3	3	3	3	3	3	21	3	1	3	2	2
4	1	1	1	1	2	22	1	2	2	3	3
5	2	2	2	2	3	23	2	3	3	1	1
6	3	3	3	3	1	24	3	1	1	2	2
7	1	1	2	3	1	25	1	3	2	1	2
8	2	2	3	1	2	26	2	1	3	2	3
9	3	3	1	2	3	27	3	2	1	3	1
10	1	1	3	2	1	28	1	3	2	2	2
11	2	2	1	3	2	29	2	1	3	3	3
12	3	3	2	1	3	30	3	2	1	1	1
13	1	2	3	1	3	31	1	3	3	3	2
14	2	3	1	2	1	32	2	1	1	1	3
15	3	1	2	3	2	33	3	2	2	2	1
16	1	2	3	2	1	34	1	3	1	2	3
17	2	3	1	3	2	35	2	1	2	3	1
18	3	1	2	1	3	36	3	2	3	1	2

* The columns 1-5 of the control orthogonal array correspond to columns 12-16 of the orthogonal array L_{36}.

Simulation Algorithm

Each row of the control orthogonal array represents a different trial design. For each trial design, the S/N ratio was evaluated using the procedure described in Sections 8.4 and 8.5. The simulation algorithm is graphically displayed in Figure 8.3. It consists of the following calculations for each row of the control orthogonal array:

1. Determine the control factor settings for a row of the control orthogonal array (Table 8.5) by using Table 8.4. For example, row 1 comprises level 1 for all

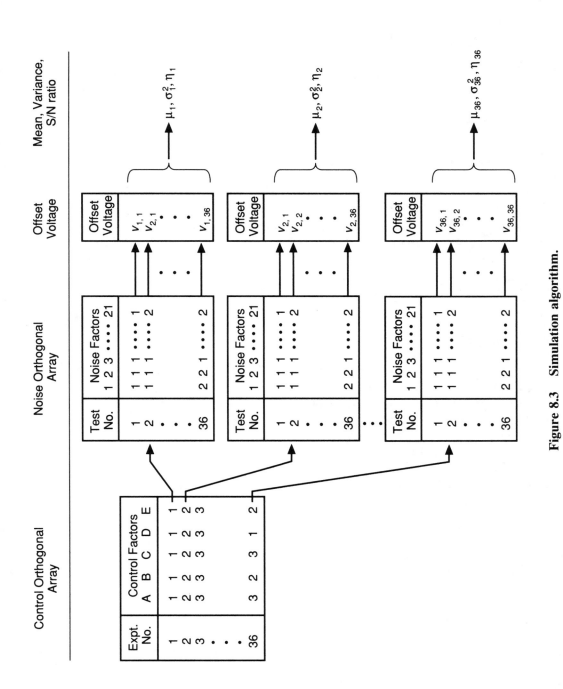

Figure 8.3   Simulation algorithm.

control factors. It corresponds to the following settings: RFM = 35.5 kΩ, RPEM = 7.5 kΩ, RNEM = 1.25 kΩ, CPCS = 10 μA, and OCS = 10 μA.

2. Use the noise orthogonal array given in Table 8.2 together with Table 8.1 to determine the 36 testing conditions. As explained earlier, these conditions simulate the effect of the noise factors in a systematic way.

3. Use the computer model for circuit analysis to calculate the offset voltage for the 36 testing conditions.

4. Calculate the mean, variance, and S/N ratio from the 36 offset voltage values.

The results of the above calculations for the 36 rows of the control orthogonal array (OA) are given in Table 8.6.

Here, the entire simulation amounts to $36 \times 36 = 1,296$ evaluations of the offset voltage. When evaluation of the response is expensive, such a large number of evaluations can prove to be impractical. Further, such a large number of evaluations is, as a rule, not necessary. Later in this chapter (Section 8.8) we discuss some practical ways of reducing the number of evaluations.

Data Analysis

Table 8.6 lists the S/N ratio and the mean offset voltage computed for each of the 36 rows of the control orthogonal array. By analyzing the S/N ratio data, we get the information in Table 8.7. The effects of the various factors on η are displayed in Figure 8.4. The 2σ confidence limits are also shown in the figure. It is clear from the plots and the ANOVA table that a major improvement in η is possible by reducing RPEM from 15 kΩ to 7.5 kΩ. A modest improvement is possible by increasing RNEM from 2.5 kΩ to 5 kΩ, and by reducing both the current sources, CPCS and OCS, from 20 μA to 10 μA. The resistance RFM has a negligible effect.

The results of the analysis of the mean offset voltage data are given in Table 8.8 and are plotted in Figure 8.4. We can see that the three resistors have only a small effect on the mean offset voltage; however, the two current sources have a large effect on the mean offset voltage. Considering that their effects on η are not large, the two current sources could be used to adjust the mean offset voltage on zero.

Optimum Settings

Considering the data analysis above, the design team chose the following two designs as potential optimum designs:

- *Optimum 1: Only change RPEM from 15 kΩ to 7.5 kΩ.* Using the procedure in Sections 8.4 and 8.5, the value of η for this design was found to be 33.58 dB compared to 29.27 dB for the starting design. In terms of the standard deviation

TABLE 8.6 MEAN AND VARIANCE OF THE OFFSET VOLTAGE

Row No. of Control OA	Mean Offset Voltage $(10^{-3}V)$	Variance of Offset Voltage $(10^{-6}V^2)$	$\eta = -10 \log_{10}$ (Variance)
1	−1.28	321	34.93
2	−3.54	1184	29.27
3	−11.47	7301	21.37
4	14.12	389	34.10
5	39.68	1789	27.47
6	−127.26	4850	23.14
7	−35.67	623	32.06
8	22.76	615	32.11
9	73.04	7737	21.11
10	−14.09	250	36.01
11	−46.93	3468	24.60
12	125.94	4685	23.29
13	70.57	1335	28.74
14	−46.56	5039	22.98
15	−26.17	763	31.17
16	−24.41	633	31.99
17	−88.82	12211	19.13
18	47.24	703	31.53
19	−6.75	3810	24.19
20	−3.77	1621	27.90
21	−1.75	336	34.74
22	−6.23	2573	25.90
23	−3.52	1135	29.45
24	−2.10	650	31.87
25	37.36	2035	26.91
26	23.43	523	32.82
27	−68.48	3534	24.52
28	−6.75	3613	24.42
29	−3.07	649	31.88
30	−2.22	775	31.11
31	−85.06	4862	23.13
32	45.88	716	31.45
33	−25.71	1089	29.63
34	69.34	6964	21.57
35	−36.44	656	31.83
36	23.27	645	31.91

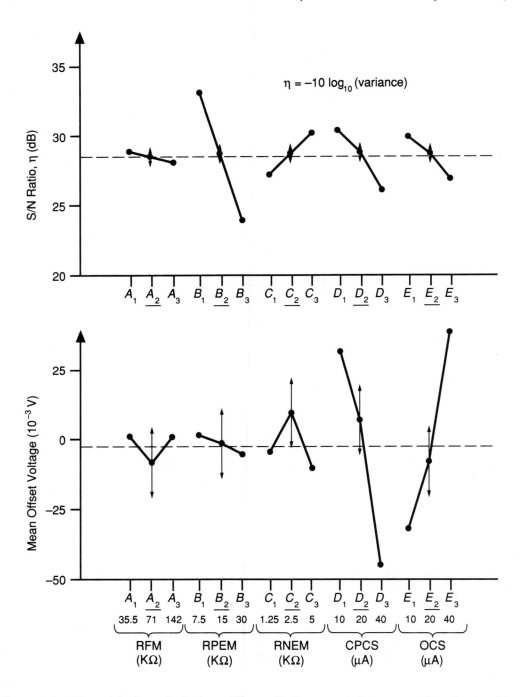

Figure 8.4 Plots of control factor effects. Underscore indicates starting level. Two-standard-deviation confidence limits are shown for the starting level.

TABLE 8.7 ANALYSIS OF S/N RATIO*

Factor	Average η by Level† 1	2	3	Degrees of Freedom	Sum of Squares	Mean Square	F
A. RFM	28.66	28.41	27.95	2	3.1	1.6	2
B. RPEM	32.87	28.45	23.70	2	504.3	252.2	348
C. RNEM	26.80	28.45	29.77	2	53.4	26.7	37
D. CPCS	30.29	28.66	26.07	2	108.2	54.1	75
E. OCS	29.63	28.61	26.78	2	50.1	25.1	35
Error				25	18.1	0.724	
Total				35	737.2	21.06	

* Overall mean η = 28.34 dB
† Starting levels are indicated by an underscore.

TABLE 8.8 ANALYSIS OF MEAN OFFSET VOLTAGE*

Factor	Average μ by Level† (10^{-3} V) 1	2	3	Degrees of Freedom	Sum of Squares (10^{-6} V²)	Mean Square (10^{-6} V²)	F
A. RFM	0.93	−8.41	0.36	2	658	329	
B. RPEM	0.84	−2.33	−5.63	2	251	126	
C. RNEM	−5.06	8.83	−10.88	2	2462	1231	2.5
D. CPCS	31.36	6.71	−45.20	2	36653	18327	37
E. OCS	−32.45	−13.63	38.97	2	32885	16443	33
Error				25	12429	497	
Total				35	85338	2438	

* Overall mean μ = −2.37 × 10^{-3} V
† Starting levels are indicated by an underscore.

of the offset voltage, this represented a reduction from 34.41 to 20.95 mV. Correspondingly, the mean offset voltage also changed from −3.54 to −1.94 mV, which is closer to the desired value of 0 mV.

- *Optimum 2: Change RPEM to 7.5 kΩ. Also change RNEM to 5.0 kΩ, CPCS to 10 μA, and OCS to 10 μA.* The η for this design was computed to be 37.0 dB, which is equivalent to a standard deviation of offset voltage equal to 14.13 mV. The mean offset voltage for this design was −16.30 mV. As discussed earlier, the mean could be easily brought to 0 mV by adjusting either CPCS or OCS, but with some adverse effect on η.

Multiple Quality Characteristics

In the discussion thus far, we have paid attention to only the dc offset voltage. Stability under ac operation is also an important consideration. For a more elaborate study of this characteristic, one must generate data similar to that in Table 8.6 and Figure 8.4. The optimum control factor setting should then be obtained by jointly considering the effects on both the dc and ac characteristics. If conflicts occur, appropriate trade-offs can be made using the quantitative knowledge of the effects. In the differential op-amp circuit example, the design team simply checked for ac stability at the two optimum conditions. For sufficient safety margin with respect to ac stability, optimum 1 was selected as the best design and was called simply *optimum design*.

This concludes the Robust Design cycle for the differential op-amp case study. For another circuit design example, see Anderson [A2]. The remaining sections of this chapter describe additional quality engineering ideas using the differential op-amp case study.

8.7 TOLERANCE DESIGN

Selecting values of control parameters that maximize the S/N ratio gives us a lower quality loss without any increase in the product cost. The optimum design arrived at in the preceding section gave a $33.58 - 29.27 = 4.31$ dB gain in S/N ratio, which is equivalent to a reduction in variance of offset voltage and, hence, the quality loss by a factor of $10^{0.431} = 2.7$. If further quality improvement is needed or desired, we must address the tolerance factors and make appropriate economic trade-offs between the increased cost of the product and the improved quality. As discussed in Chapter 2, for best engineering economics, tolerance design should come only after the S/N ratio has been maximized.

The first step in tolerance design is to determine the contribution of the noise factor to the quality loss. This can be done by using the procedure in Section 8.4. Table 8.9 gives the contribution to total variance of the noise factors for the optimum design, as well as the starting design. Note that the contribution to the total variance by a noise factor is computed by dividing the sum of squares due to that factor by the degrees of freedom for that total sum of squares, namely 35. In this table, we list only the top four contributors. Together, these four noise factors account for more than 95 percent of the total variance. Thus, the contribution of noise factors exhibits the typical Pareto principle. To improve the joint economics of product cost and quality loss, we should consider the following issues:

1. For noise factors that contribute a large amount to the variance of the offset voltage, consider ways of reducing their variation. This is the situation with the noise factors SIENP, AFNO, AFNM, and SIEPP.

2. For the noise factors that contribute only a small amount to the variance, consider ways of saving cost by allowing more variation. This is the situation with the noise factors other than those listed above.

TABLE 8.9 CONTRIBUTION TO THE VARIANCE OF THE OFFSET VOLTAGE BY VARIOUS NOISE FACTORS

Noise Factor	Contribution to the Variance of Offset Voltage (10^{-6} V^2)	
	Starting Design	Optimum Design
SIENP	687	232
AFNO	224	76
AFNM	167	57
SIEPP	53	55
Remainder	54	19
Total	1184	439

Here, we use some fictitious cost values to illustrate the economic trade-off involved in tolerance design. Suppose there are three layout techniques for the integrated circuit. We call technique 1 the baseline technique. If technique 2 is used, it leads to a factor of two reduction in variance of SIENP, and it costs $0.20 more per circuit to manufacture. If technique 3 is used, it leads to a factor of four reduction in variance of SIENP, and it costs $1.00 more per circuit to manufacture. Which layout technique should we specify? The layout technique is a tolerance factor because it affects both the quality loss and unit manufacturing cost. Selecting levels of such factors is the goal of tolerance design.

To select the layout technique, we need to know the quality loss coefficient, which was defined in Chapter 2. Suppose the LD5O point (the point at which one-half of the circuits fail) is $\Delta_0 = 100$ mV and the cost of failure is $A_0 = \$30$. Then, the quality loss coefficient, k, is given by

$$k = \frac{A_0}{\Delta_0^2} = \frac{30}{100^2} = 0.003 \quad \text{dollars}/(\text{mV})^2 .$$

So, the quality loss due to variation in SIENP when layout technique 1 is used with the optimum design is

$$Q' = k\,\sigma^2 = 0.003 \times 232 = \$0.70/\text{circuit}.$$

Since the variance of offset voltage caused by SIENP is directly proportional to its own variance, the quality loss under techniques 2 and 3 when optimum design is used would be $\$0.35$/circuit and $\$0.18$/circuit, respectively. In each case, the quality loss due to the other noise factors is $\$0.62$/circuit. Consequently, the total cost (which consists of product cost plus the quality loss due to all noise factors) for layout techniques 1, 2, and 3 would be $\$1.32$/circuit, $\$1.17$/circuit, and $\$1.80$/circuit, respectively. These costs are listed in Table 8.10. It is clear that in order to minimize the total cost incurred by the customer (which is the quality loss) and by the manufacturer (extra cost for the layout technique), we should choose technique 2 for layout. Cost trade-offs of this nature should be done with all tolerance factors.

TABLE 8.10 TOLERANCE DESIGN—SELECTION OF LAYOUT TECHNIQUE

Layout Technique	Incremental Product ($/circuit)	Quality Loss ($/circuit)		Total Cost* ($/circuit)
		Due to Variation in SIENP	Due to Variation in Other Noise Factors	
Technique 1	0	0.70	0.62	1.32
Technique 2	0.20	0.35	0.62	1.17
Technique 3	1.00	0.18	0.62	1.80

* Total cost is the sum of incremental product cost and quality loss due to all sources.

The total economics of performing Robust Design followed by tolerance design are summarized in Table 8.11. For the starting design, the quality loss was $3.55/circuit. After design optimization, the quality loss reduced to $1.32/circuit, without any increase in manufacturing cost. Through tolerance design, we can further reduce the quality loss by $0.35/circuit by spending an extra $0.20/circuit on a better layout technique. If we skipped the design optimization step (finding the best levels of the control factors), we would need to use the most expensive layout technique. Compared to the layout technique 1, this represents a $0.54/circuit saving in total cost. However, compared to the optimum design with layout technique 2, it represents an extra total cost of $1.84/circuit! This is the economic advantage of first reducing the sensitivity to noise factors and then controlling the noise factors that are the causes of variation.

TABLE 8.11 ANALYSIS OF COST SAVING

Design Type	Quality Loss ($/circuit)	Product Cost ($/circuit)	Total Cost ($/circuit)
1. Starting design with layout technique 1	3.55	0	3.55
2. Optimum design with layout technique 1	1.32	0	1.32
3. Optimum design with layout technique 2	0.97	0.20	1.17
4. Starting design with layout technique 3 (obtained by performing only tolerance design)	2.01	1.00	3.01

8.8 REDUCING THE SIMULATION EFFORT

It often proves to be very expensive to simulate the product performance even under one set of parameter values. In that case, using an array such as L_{36} to evaluate the S/N ratio for each combination of control factor settings in the control orthogonal array can be prohibitively expensive. This section presents some practical approaches for reducing the number of testing conditions used in evaluating the S/N ratio for each combination of control factor levels.

Selecting Major Noise Factors

By comparing the two columns in Table 8.9 that correspond to the starting and optimum designs, we notice that the sensitivity to all noise factors is reduced more or less uniformly by a factor of about 3. (Tolerance in SIEPP is an exception, but its contribution to the total variance is rather small.)

Similar breakdowns of the variance of the offset voltage corresponding to several combinations of control factor settings lead to the conclusion that the same four noise factors account for most of the variance of the offset voltage. This implies that it would be adequate to include only a few major noise sources in a Robust Design project. Taguchi and other researchers have observed in many areas of engineering design that if we reduce the sensitivity to a few major noise factors, the sensitivity to other noise factors also reduces by a similar amount. Thus, we can greatly reduce the size of experimental or simulation effort in a Robust Design project by considering only a small number of important noise factors. Often, engineering judgment can tell us what the important noise factors are. In other situations, an orthogonal array experiment similar to that in Section 8.4 can be used to verify our judgment or to identify important noise factors.

For the circuit, it would have been adequate to consider only the tolerances in the parameters SIENP, AFNO, and AFNM for optimizing the circuit for offset voltage. This would mean that for the purpose of optimization the L_9 array could be used in place of the L_{36} array for simulating the effect of noise factors. This amounts to a 4-fold reduction in the total number of evaluations of the offset voltage.

Compound Noise Factor

Further reduction in the simulation effort is possible by compounding noise factors. Suppose we can establish through small simulations that the directionality of the effects of the noise factors on the response is consistent. This means, for example, establishing that when SIENP is increased, the offset voltage consistently increases or decreases regardless of the values of the other parameters. Suppose we have established a similar directionality for the other two major noise factors—that is, AFNO and AFNM. Then, we construct a *compound noise factor*, called CN, as follows:

- *Level 1 of* CN. Select the levels for SIENP, AFNO, and AFNM so that we get a compounded effect of lowering the offset voltage—that is, pushing it to the negative side. From Table 8.3, we see that level 1 of CN should consist of level 1 of SIENP, level 3 of AFNO, and level 1 of AFNM.

- *Level 2 of* CN. Take nominal levels for SIENP, AFNO, and AFNM.

- *Level 3 of* CN. Select the levels of SIENP, AFNO, and AFNM so that we get a compounded effect of increasing the offset voltage—that is, pushing it to the positive side. Again from Table 8.3, we see that level 3 of CN should consist of level 3 of SIENP, level 1 of AFNO, and level 3 of AFNM.

The three levels of the compound noise factor are simply three testing conditions. Thus, for each combination of control factor settings, we would evaluate the offset voltage at only three testing conditions and compute the S/N ratio from these three values. This reduces the testing conditions to bare bones. In order for the optimization performed with the compound noise factor to give meaningful results, it is necessary that we know the following:

1. What the major noise factors are.

2. The directionality of their effects on the chosen quality characteristic, and

3. That the directionality of these effects does not depend on the settings of the control factors.

Note that if conditions two and three above are violated, then during simulation or experimentation the effect of one noise factor may get compensated for by another noise factor. In that case, the optimization based on compounded noise can give confused results.

In any given project, one can use a large orthogonal array like L_{36} under a starting design to identify major noise factors and the directionality of their effects. Properly selecting the quality characteristic helps ensure consistency in the directionality of the effects of noise factors. Lack of such consistency can only be detected through the final verification experiment.

Is the S/N ratio calculated from the compound noise factor equal to the S/N ratio calculated from an orthogonal array based simulation? Of course, they need not be equal. However, for optimization purposes, it does not make any difference whether we use one or the other, except for the computational effort.

8.9 ANALYSIS OF NONLINEARITY

It is helpful to examine the nature of the nonlinearity of the relationship between the circuit parameters and the offset voltage. Figure 8.5 shows the plot of offset voltage as a function of SIENP/SIENM for RPEM = 15 kΩ and RPEM = 7.5 kΩ. Note that the nominal value of SIENP is SIENM. Therefore, we take the ratio SIENP/SIENM to study the sensitivity of offset voltage to variation in SIENP. All other circuit parameters are set at their nominal levels for these plots. It is clear that sensitivity to variation in SIENP/SIENM is altered by changing another parameter, that is, RPEM. Compare this with the nonlinearity studied in Chapter 2, Section 2.5 where a change in the nominal value of the gain leads to a change in the sensitivity to variation in the gain itself. In Robust Design, we are interested in exploiting both the types of nonlinearities to reduce successfully the sensitivity to noise factors.

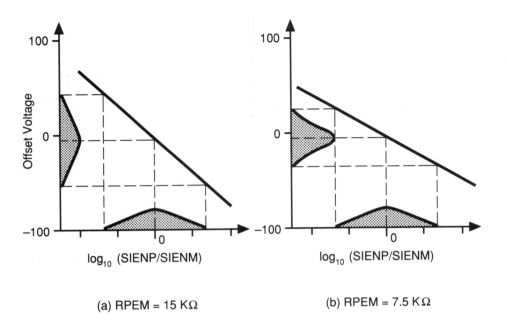

(a) RPEM = 15 KΩ   (b) RPEM = 7.5 KΩ

Figure 8.5   Relationship between offset voltage and SIENP when (a) RPEM = 15 kΩ and (b) RPEM = 7.5 kΩ are shown in the two plots. Note that changing RPEM changes the slope. Also shown in the plots are the transfer of variance from SIEPM to offset voltage.

8.10 SELECTING AN APPROPRIATE S/N RATIO

One benefit of using an orthogonal array with many degrees of freedom for error is that we can judge the additivity of the factor effects. Referring to Table 8.7, we see that in this case study the error mean square for the chosen S/N ratio (signed-target) is 0.724. The total degrees of freedom for the control factors is 10, the corresponding sum of squares is 3.1 + 504.3 + 53.4 + 108.2 + 50.1 = 719.1 so that the mean square for the factor effects is 71.91. Thus, the ratio of the error mean square (0.724) to the mean square for the factor effects is 0.01. This implies that the additivity of the factor effects is excellent for the chosen S/N ratio.

What if we had taken the absolute value of the offset voltage and treated it as a smaller-the-better type quality characteristic? ANOVA was performed by using the corresponding S/N ratio, which yielded the error mean square equal to 9.26 and the mean square for the factor effects equal to 70.21. For this S/N ratio, the ratio of the error mean square to the mean square for the factor effects is 0.13, which is more than 10 times larger than the corresponding ratio for the chosen S/N ratio. This is evidence that the chosen S/N ratio, namely, the signed-target S/N ratio, is substantially preferable to the smaller-the-better type S/N ratio for this case study. Thus, when in doubt, one can conduct ANOVA with the candidate S/N ratios and pick the one that gives the smallest relative error mean square.

Some caution is needed in using the approach of this section for selecting the appropriate S/N ratio. The candidate S/N ratios should only be those that can be justified from engineering knowledge. The orthogonal array can then help identify which adjustments are easier to accomplish. The approach of this section can also be used in deciding which is a better quality characteristic from among a few that can be justified from engineering knowledge.

8.11 SUMMARY

- Computer models can be used in place of the hardware experiments for optimizing a product or process design. The eight Robust Design steps described in Chapter 4 are valid equally well with computer models. Some of these steps can be automated with the help of appropriate software, thus making it easier to optimize the design.

- The differential op-amp example helps illustrate:

    — How to optimize a product design, particularly when a computer model is available.

    — How an orthogonal array can be used to simulate the variation in noise factors and, thus, estimate the quality loss and manufacturing yield.

    — How an orthogonal array can be used to select the most appropriate S/N ratio from among several candidates.

    — How to perform tolerance design.

- Three common methods of evaluating the mean and variance of a product's response resulting from variations in many noise factors are Monte Carlo simulation, Taylor series expansion, and orthogonal array based simulation. In the orthogonal array based simulation, proposed by Genichi Taguchi, we use an orthogonal array to sample the domain of noise factors. For each noise factor, we take either two levels ($\mu - \sigma$, $\mu + \sigma$) or three levels ($\mu - \sqrt{3/2}\ \sigma$, μ, $\mu + \sqrt{3/2}\ \sigma$).

- The benefits of the orthogonal array based simulation method are:

    — It gives common testing conditions for comparing two or more combinations of control factor levels.

    — With the help of ANOVA, it can be used to estimate the contribution of each noise factor.

— Compared to the Monte Carlo method, it needs a much smaller number of testing conditions, without sacrificing any accuracy.

— When interactions and correlations among the noise factors are strong, it gives more accurate estimates of mean and variance compared to the Taylor series expansion.

• Orthogonal array based simulation of noise factors can be used to evaluate the S/N ratio for each combination of control factor settings used in a matrix experiment. However, this may prove to be too expensive in some projects. Selecting only major noise factors and forming a compound noise factor can greatly reduce the simulation effort and, hence, are recommended for design optimization.

• For the differential op-amp circuit example, the quality characteristic (offset voltage) was of the signed-target type. It had five control factors and 21 noise factors. The standard orthogonal array L_{36} was used to simulate the effect of noise factors and also to construct trial combinations of control factor settings. Thus, $36 \times 36 = 1,296$ circuit evaluations were performed. The selected optimum control factor settings gave a 4.31 dB increase in S/N ratio. This represents a 63 percent reduction in the mean square offset voltage.

• Selecting the levels of control factors that maximize the S/N ratio gives a lower quality loss without any increase in product cost. Thus, here, 63 percent reduction in quality loss was obtained. If further quality improvement is needed or desired, tolerance factors must be addressed and appropriate economic trade-offs must be made between the increased cost of the product and the improved quality. Here, a 11.4 percent reduction in total cost was achieved by tolerance design. For best engineering economics (as noted in Chapter 2), tolerance design should come only after the S/N ratio has been maximized (see Table 8.11).

• The first step of tolerance design is to determine the contribution of each noise factor to the quality loss. To improve the joint economics of product cost and quality loss, one should consider two issues:

   1. Ways of reducing the variation of the noise factors that contribute a large amount to the quality loss

   2. Ways of saving cost by allowing wider variation for the noise factors that contribute only a small amount to the quality loss

• The differential op-amp example illustrates an interesting nonlinearity among the response (offset voltage) and the circuit parameters. The sensitivity of the offset voltage to the noise factor SIENP (particularly the ratio SIENP/SIENM) is altered by changing another parameter, namely RPEM. It is useful to compare this with the nonlinearity studied in Chapter 2, Section 2.5 where the sensitivity to the variation in gain is altered by changing the nominal value of the gain itself. In Robust Design, both types of nonlinearities are used for reducing sensitivity to noise factors.

- Matrix experiments using orthogonal arrays are useful for selecting the most appropriate S/N ratio from among a few candidates. In such an application, a few degrees of freedom should be reserved for estimating the error variance. A smaller error variance, relative to the mean square for the control factor effects, signifies that the additivity of the particular S/N ratio is better, and, hence, the S/N ratio is more suitable.

Chapter 9

DESIGN OF
DYNAMIC SYSTEMS

Dynamic systems are those in which we want the system's response to follow the levels of the signal factor in a prescribed manner. Chapter 5 gave several examples of dynamic systems and the corresponding S/N ratios. The changing nature of the levels of the signal factor and the response make designing a dynamic system more complicated than designing a static system. However, the eight steps of Robust Design described in Chapter 4 are still valid. This chapter describes the design of a temperature control circuit to illustrate the application of the Robust Design method to a dynamic system. This chapter has six sections:

- Section 9.1 describes the temperature control circuit and its main function (Step 1 of the Robust Design steps described in Chapter 4).

- Section 9.2 gives the signal, control, and noise factors for the circuit (Steps 2 and 4).

- Section 9.3 discusses the selection of the quality characteristics and the S/N ratios (Step 3).

- Section 9.4 describes the steps in the optimization of the circuit, including verification of the optimum conditions (Steps 5 through 8).

- Section 9.5 discusses iterative optimization using matrix experiments based on orthogonal arrays.

- Section 9.6 summarizes the important points of this chapter.

9.1 TEMPERATURE CONTROL CIRCUIT AND ITS FUNCTION

The function of a temperature controller is to maintain the temperature of a room, a bath, or some object at a target value. The heating and cooling systems in a house are common examples of temperature controllers. From the Robust Design point of view, a temperature controller can be divided into three main modules (see Figure 9.1): (1) temperature sensor, (2) temperature control circuit, and (3) heating (or cooling) element. The function of the temperature sensor is to measure the temperature accurately and pass that information to the temperature control circuit. The temperature control circuit provides a way of setting the target temperature. It also compares the observed temperature with the target temperature and makes a decision about turning the heater ON or OFF. Control systems where the heater can be only ON or OFF are called *bang-bang controllers*. In some other types of controllers, the amount of heat delivered can also be varied. Here, we discuss the design of a temperature control circuit for a bang-bang controller.

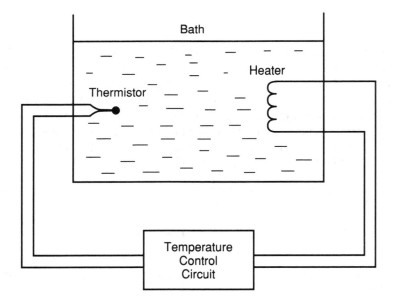

Figure 9.1　Block diagram of a temperature control system.

　For a particular target temperature, the circuit must turn a heater ON or OFF to control the heat input, which makes the temperature controller a dynamic problem. Further, the target temperature can be changed by the users (that is, on one day users may want the target temperature at 80°C, whereas on another day they may want it at 90°C) which also makes the design of a temperature control circuit a dynamic problem. Thus, designing a temperature control circuit is a doubly dynamic problem.

Figure 9.2 shows a standard temperature control circuit. (The tolerance design of a slightly modified version of this circuit was discussed by Akira Tomishima [T9].) Suppose we want to use the circuit to maintain the temperature of a bath at a value that is above the ambient temperature. The temperature of the bath is sensed by a thermistor, which we assume to have a negative temperature coefficient; that is, as shown in Figure 9.3, the thermistor resistance, R_T, decreases with an increase in the temperature of the bath. When the bath temperature rises above a certain value, the resistance R_T drops below a threshold value so that the difference in the voltages between terminals 1 and 2 of the amplifier becomes negative. This actuates the relay and turns OFF the heater. Likewise, when the temperature falls below a certain value, the difference in voltages between the terminals 1 and 2 becomes positive so that the relay is actuated and the heater is turned ON. In the terminology of Chapter 5, bang-bang controllers are Continuous-Discrete (C-D) type dynamic problems.

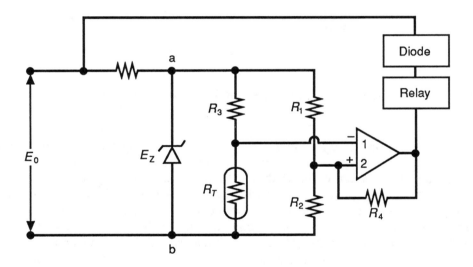

Figure 9.2   Temperature control circuit.

Let us denote the value of the thermistor resistance at which the heater turns ON, by R_{T-ON} and the value of the thermistor resistance at which the heater turns OFF by R_{T-OFF}. Consider the operation of the controller for a particular target temperature. The values of R_{T-ON} and R_{T-OFF} can change due to variation in the values of the various circuit components. This is graphically displayed in Figure 9.4. In the figure, m_T is the thermistor resistance at the target temperature; and m_{ON} and m_{OFF} are the mean values of R_{T-ON} and R_{T-OFF}, respectively.

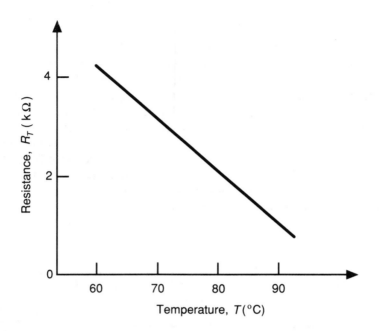

Figure 9.3   **Resistance vs. temperature plot for a thermistor with negative coefficient.**

The separation of m_{ON} and m_{OFF} from m_T is commonly called *hysteresis*. It allows a period of time for the heater to stay ON and the temperature to rise slightly above the target temperature during the ON part of the heating cycle. Similarly, in the OFF part of the heating cycle, it allows the heater to stay OFF for a period of time and the temperature to drop a little bit below the target temperature. The heat transfer parameters of the bath (such as the thermal capacity of the bath, the rate at which heat is input by the heater, and the rate at which heat is lost to the surroundings) and the considerations of longevity of the heater (too frequent transitions of ON and OFF can greatly reduce the heater life) dictate the needed hysteresis for the temperature control circuit. Of course, the quality loss resulting from the variation of the temperature around the target is also important in deciding the amount of hysteresis.

Thus, the ideal function of the temperature control circuit can be written as follows:

1. Provides a mechanism to set the target value m_T of R_T.

2. For every target value of R_T, it turns the heater ON and OFF at the corresponding specified values of R_{T-ON} and R_{T-OFF}.

This is clearly a doubly dynamic problem—first for the ability to set m_T (C-C type system), second for the ON-OFF transition (C-D type system).

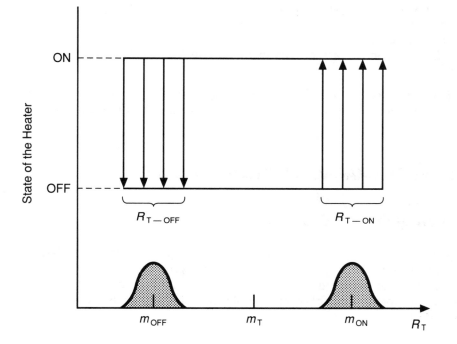

Figure 9.4   Hysteresis and variation in the operation of a temperature control circuit.

9.2 SIGNAL, CONTROL, AND NOISE FACTORS

Let us first examine the choice of a signal factor. Referring to the circuit diagram of Figure 9.2, we notice that the four resistances, R_1, R_2, R_3 and R_T, form a Wheatstone bridge. Therefore, any one of the three resistances, R_1, R_2 or R_3, can be used to adjust the value of R_T at which the bridge balances. We decide to use R_3, and, thus, it is our signal factor for deciding the temperature setting. The resistance R_T by itself is the signal factor for the ON-OFF operations.

The purpose of the Zener diode (nominal voltage = E_z) in the circuit is to regulate the voltage across the terminals a and b (see Figure 9.2). That is, when the Zener diode is used, the voltage across the terminals a and b remains constant even if the power supply voltage, E_0, drifts or fluctuates. Thus it reduces the dependence of the threshold values R_{T-ON} and R_{T-OFF} on the power supply voltage E_0.

As a general rule, the nominal values of the various circuit parameters are potential control factors, except for those completely defined by the tracking rules. In the temperature control circuit, the control factors are R_1, R_2, R_4 and E_z. Note that we do not take E_0 as a control factor. As a rule, the design engineer has to make the decision about which parameters should be considered control factors and which should not. The main function of E_0 is to provide power for the operation of the relay, and its nominal value is not as important for the ON-OFF operation. Hence, we do not include E_0 as a control factor. The tolerances on R_1, R_2, R_4, E_z, and E_0 are the noise factors.

For proper operation of the circuit, we must have $E_z < E_0$. Also, R_4 must be much bigger than R_1 or R_2. These are the tracking relationships among the circuit parameters.

9.3 QUALITY CHARACTERISTICS AND S/N RATIOS

Selection of Quality Characteristics

The resistances R_{T-ON} and R_{T-OFF} are continuous variables that are obviously directly related to the ON-OFF operations; together, they completely characterize the circuit function. Through standard techniques of circuit analysis, one can express the values of R_{T-ON} and R_{T-OFF} as following simple mathematical functions of the other circuit parameters:

$$R_{T-ON} = \frac{R_3 R_2 (E_z R_4 + E_0 R_1)}{R_1 (E_z R_2 + E_z R_4 - E_0 R_2)} \tag{9.1}$$

$$R_{T-OFF} = \frac{R_3 R_2 R_4}{R_1 (R_2 + R_4)} \ . \tag{9.2}$$

Thus, by the criteria defined in Chapter 6, R_{T-ON} and R_{T-OFF} are appropriate choices for the quality characteristics.

Suppose Equations (9.1) and (9.2) for the evaluation of R_{T-ON} and R_{T-OFF} were not available and that hardware experiments were needed to determine their values. Measuring R_{T-ON} and R_{T-OFF} would still be easy. It could be accomplished by incrementing the values of R_T through small steps until the heater turns ON and decrementing the values of R_T until the heater turns OFF.

Selection of S/N Ratio

The ideal relationship of R_{T-ON} and R_{T-OFF} with R_3 (the signal factor) is linear, passing through the origin, as shown in Figure 9.5. So for both quality characteristics, the appropriate S/N ratio is the C-C type S/N ratio, described in Chapter 5. Suppose for some particular levels of the control factors and particular tolerances associated with

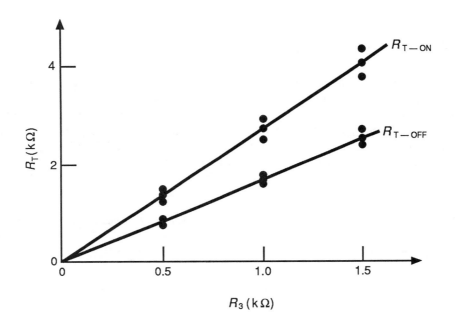

Figure 9.5   Plot of R_{T-ON} and R_{T-OFF} vs. R_3 for the starting design.

the noise factors we express the dependence of R_{T-ON} on R_3 by the following equation obtained by the least squares fit:

$$R_{T-ON} = \beta R_3 + e \tag{9.3}$$

where β is the slope and e is the error. Note that any nonlinear terms in R_3 (such as R_3^2 or R_3^3) are included in the error e. The S/N ratio for R_{T-ON} is given by

$$\eta = 10 \log_{10} \frac{\beta^2}{\sigma_e^2}. \tag{9.4}$$

Similarly, let the dependence of R_{T-OFF} on R_3 be given by

$$R_{T-OFF} = \beta' R_3 + e' \tag{9.5}$$

where β' is the slope and e' is the error. Then, the corresponding S/N ratio for R_{T-OFF} is given by

$$\eta' = 10 \log_{10} \frac{\beta'^2}{\sigma_{e'}^2} . \qquad (9.6)$$

Evaluation of the S/N Ratio

Let us first see the computation of the S/N ratio for R_{T-ON}. The nominal values of the circuit parameters under the starting conditions, their tolerances (three-standard-deviation limits), and the three levels for testing are shown in Table 9.1. These levels were computed by the procedure described in Chapter 8; that is, for each noise factor, the levels 1 and 3 are displaced from level 2, which is equal to its mean value, on either side by $\sqrt{3/2}\ \sigma$, where σ is one-third the tolerance.

TABLE 9.1 NOISE AND SIGNAL FACTORS FOR TEMPERATURE CONTROL CIRCUIT

Factor	Mean*	Tolerance (%)	Levels (Multiply by Mean for Noise Factors)		
			1	2	3
A. R_1	4.0 kΩ	5	1.0204	1.0	0.9796
B. R_2	8.0 kΩ	5	1.0204	1.0	0.9796
C. R_4	40.0 kΩ	5	1.0204	1.0	0.9796
D. E_0	10.0 V	5	1.0204	1.0	0.9796
F. E_z	6.0 V	5	1.0204	1.0	0.9796
M. R_3 (signal)	–	–	0.5 kΩ	1.0 kΩ	1.5 kΩ

* Mean values listed here correspond to the nominal values for the starting design.

The ideal relationship between R_3 and R_{T-ON} is a straight line through the origin with the desired slope. Second- and higher-order terms in the relationship between R_3 and R_{T-ON} should be, therefore, minimized. Thus, we take three levels for the signal factor (R_3): 0.5 kΩ, 1.0 kΩ, and 1.5 kΩ. Here R_{T-ON} must be zero when R_3 is zero. So, with three levels of R_3, we can estimate the first-, second-, and third-order terms in the dependence of R_{T-ON} on R_3. The first order, or the linear effect, constitutes the desired signal factor effect. We include the higher-order effects in the noise variance so they are reduced with the maximization of η. [It is obvious from Equations (9.1) and (9.2) that the second- and higher-order terms in R_3 do not appear in this circuit. Thus, taking only one level of R_3 would have been sufficient. However, we take three levels to illustrate the general procedure for computing the S/N ratio.]

As discussed in Chapter 8, an orthogonal array (called *noise orthogonal array*) can be used to simulate the variation in the noise factors. In addition to assigning noise factors to the columns of an orthogonal array, we can also assign one of the columns to the signal factor. From the values of R_{T-ON} corresponding to each row of the noise orthogonal array, we can perform least squares regression (see Section 5.4, or Hogg and Craig [H3], or Draper and Smith [D4]) to estimate β and σ_e^2 and then the S/N ratio, η.

Chapter 8 pointed out that the computational effort can be reduced greatly by forming a *compound noise factor*. For that purpose, we must first find directionality of the changes in R_{T-ON} caused by the various noise factors. By studying the derivatives of R_{T-ON} with respect to the various circuit parameters, we observed the following relationships: R_{T-ON} increases whenever R_1 decreases, R_2 increases, R_4 decreases, E_0 increases, or E_z decreases. (If the formula for R_{T-ON} were complicated, we could have used the noise orthogonal array to determine the directionalities of the effects.) Thus, we form the three levels of the compound noise factor as follows:

$$(CN)_1: \quad (R_1)_3, \; (R_2)_1, \; (R_4)_3, \; (E_0)_1, \; (E_z)_3$$

$$(CN)_2: \quad (R_1)_2, \; (R_2)_2, \; (R_4)_2, \; (E_0)_2, \; (E_z)_2$$

$$(CN)_3: \quad (R_1)_1, \; (R_2)_3, \; (R_4)_1, \; (E_0)_3, \; (E_z)_1 \; .$$

For every level of the signal factor, we calculate R_{T-ON} with the noise factor levels set at the levels $(CN)_1$, $(CN)_2$, and $(CN)_3$. Thus, we have nine testing conditions for the computation of the S/N ratio. The nine values of R_{T-ON} corresponding to the starting values of the control factors ($R_1 = 4.0 \; k\Omega$, $R_2 = 8.0 \; k\Omega$, $R_4 = 40.0 \; k\Omega$, and $E_z = 6.0 \; k\Omega$) are tabulated in Table 9.2. Let y_i denote the value of R_{T-ON} for the ith testing condition; and let $R_{3(i)}$ be the corresponding value of R_3. Then, from standard least squares regression analysis (see Section 5.4), we obtain

$$\beta = \frac{R_{3(1)}y_1 + R_{3(2)}y_2 + \cdots + R_{3(9)}y_9}{R_{3(1)}^2 + R_{3(2)}^2 + \cdots + R_{3(9)}^2} \; . \tag{9.7}$$

The error variance is given by

$$\sigma_e^2 = \frac{1}{8} \sum_{i=1}^{9} (y_i - \beta R_{3(i)})^2 \; . \tag{9.8}$$

Substituting the appropriate values from Table 9.2 in Equations (9.7) and (9.8) we obtain, $\beta = 2.6991$ and $\sigma_e^2 = 0.030107$. Thus, the S/N ratio for R_{T-ON} corresponding to the starting levels of the control factors is

$$\eta = 10 \log_{10} \frac{\beta^2}{\sigma_e^2} = 23.84 \ dB.$$

The S/N ratio, η', for R_{T-OFF} can be computed in exactly the same manner as we computed the S/N ratio for R_{T-ON}.

Note that for dynamic systems, one must identify the signal factor and define the S/N ratio before making a proper choice of testing conditions. This is the case with the temperature control circuit.

TABLE 9.2 TESTING CONDITIONS FOR EVALUATING η

Test No.	R_3 (Signal Factor) (kΩ)	CN (Compound Noise Factor)	$y = R_{T-ON}$* (kΩ)
1	0.5	$(CN)_1$	1.2586
2	0.5	$(CN)_2$	1.3462
3	0.5	$(CN)_3$	1.4440
4	1.0	$(CN)_1$	2.5171
5	1.0	$(CN)_2$	2.6923
6	1.0	$(CN)_3$	2.8880
7	1.5	$(CN)_1$	3.7757
8	1.5	$(CN)_2$	4.0385
9	1.5	$(CN)_3$	4.3319

* These R_{T-ON} values correspond to the starting levels for the control factors.

9.4 OPTIMIZATION OF THE DESIGN

Control Factors and Their Levels

The four control factors, their starting levels, and the alternate levels are listed in Table 9.3. For the three resistances (R_1, R_2, and R_4), level 2 is the starting level, level 3 is 1.5 times the starting level, and level 1 is $1/1.5 = 0.667$ times the starting level. Thus, we include a fairly wide range of values with the three levels for each control factor. Since the available range for E_z is restricted, we take its levels somewhat closer. Level 3 of E_z is 1.2 times level 2, while level 1 is 0.8 times level 2.

TABLE 9.3 CONTROL FACTORS FOR THE TEMPERATURE CONTROL CIRCUIT*

Factor	Levels*		
	1	**2**	**3**
A. R_1 (kΩ)	2.668	4.0	6.0
B. R_2 (kΩ)	5.336	8.0	12.0
C. R_4 (kΩ)	26.68	40.0	60.0
D. E_z (V)	4.8	6.0	7.2

* Starting levels are indicated by an underscore.

Control Orthogonal Array

The orthogonal array L_9, which has four 3-level columns, is just right for studying the effects of the four control factors. However, by taking a larger array, we can also get a better indication of the additivity of the control factor effects. Further, computation is very inexpensive for this circuit. So, we use the L_{18} array to construct the *control orthogonal array*. The L_{18} array and the assignment of the control factors to the columns are given in Table 9.4. The control orthogonal array for this study is the submatrix of L_{18} formed by the columns assigned to the four control factors.

Data Analysis and Optimum Conditions

For each row of the control orthogonal array, we computed β and η for R_{T-ON}. The values of η and β^2 are shown in Table 9.4 along with the control orthogonal array. The possible range for the values of β is 0 to ∞ and we are able to get a better additive model for β in the log transform. Therefore, we study the values of β in the decibel scale, namely $20 \log_{10} \beta$. The results of performing the analysis of variance on η and $20 \log_{10} \beta$ are tabulated in Tables 9.5 and 9.6. The control factor effects on η and $20 \log_{10} \beta$ are plotted in Figure 9.6(a) and (b). Also shown in the figure are the control factor effects on η' and $20 \log_{10} \beta'$ corresponding to the quality characteristic R_{T-OFF}. The following observations can be made from Figure 9.6(a):

- For the ranges of control factor values listed in Table 9.3, the overall S/N ratio for the OFF function is higher than that for the ON function. This implies that the spread of R_{T-ON} values caused by the noise factors is wider than the spread of R_{T-OFF} values.

- The effects of the control factors on η' are much smaller than the effects on η.
- R_1 has negligible effect on η or η'.
- η can be increased by decreasing R_2; however, it leads to a small reduction in η'.
- η can be increased by increasing R_4; however, it too leads to a small reduction in η'.
- η can be increased by increasing E_z, with no adverse effect on η'.

TABLE 9.4 CONTROL ORTHOGONAL ARRAY AND DATA FOR R_{T-ON}

Expt. No.	Column Numbers and Factor Assignment*								η (dB)	β^2
	1 e	2 e	3 A	4 B	5 C	6 e	7 D	8 e		
1	1	1	1	1	1	1	1	1	22.41	9.59
2	1	1	2	2	2	2	2	2	23.84	7.29
3	1	1	3	3	3	3	3	3	24.79	6.12
4	1	2	1	1	2	2	3	3	25.85	5.33
5	1	2	2	2	3	3	1	1	24.19	7.12
6	1	2	3	3	1	1	2	2	19.47	15.66
7	1	3	1	2	1	3	2	3	22.25	19.27
8	1	3	2	3	2	1	3	1	23.61	15.04
9	1	3	3	1	3	2	1	2	24.93	1.42
10	2	1	1	3	3	2	2	1	24.23	31.22
11	2	1	2	1	1	3	3	2	24.50	3.07
12	2	1	3	2	2	1	1	3	22.13	5.03
13	2	2	1	2	3	1	3	2	26.02	11.31
14	2	2	2	3	1	2	1	3	16.19	61.02
15	2	2	3	1	2	3	2	1	24.60	1.49
16	2	3	1	3	2	3	1	2	20.26	58.36
17	2	3	2	1	3	1	2	3	25.94	2.49
18	2	3	3	2	1	2	3	1	23.05	3.95

* Empty column is indicated by e.

TABLE 9.5 ANALYSIS OF S/N RATIO DATA FOR R_{T-ON}*

Factor	Average η by Level† 1	2	3	Degrees of Freedom	Sum of Squares	Mean Square	F
A. R_1	23.50	23.04	23.16	2	0.7	0.4	–
B. R_2	24.70	23.58	21.42	2	33.33	16.67	22
C. R_4	21.31	23.38	25.02	2	41.40	20.70	27
D. E_z	21.68	23.39	24.64	2	26.37	13.19	17
Error				9	6.87	0.76	
Total				17	108.67		

* Overall mean η = 23.23 dB.
† Starting levels are indicated by an underscore.

TABLE 9.6 ANALYSIS OF $20 \log_{10} \beta$ FOR R_{T-ON}*

Factor	Average $20 \log_{10}\beta$ by Level† 1	2	3	Degrees of Freedom	Sum of Squares	Mean Square	F
A. R_1	12.18	9.27	6.01	2	114.2	57.1	94
B. R_2	4.86	8.92	13.68	2	233.5	116.8	191
C. R_4	10.55	9.01	7.89	2	21.4	10.7	18
D. E_z	10.40	9.01	8.05	2	16.8	8.4	14
Error				9	5.5	0.61	
Total				17	391.4		

* Overall mean value of $20 \log_{10}\beta$ = 9.15 dB.
† Starting levels are indicated by an underscore.

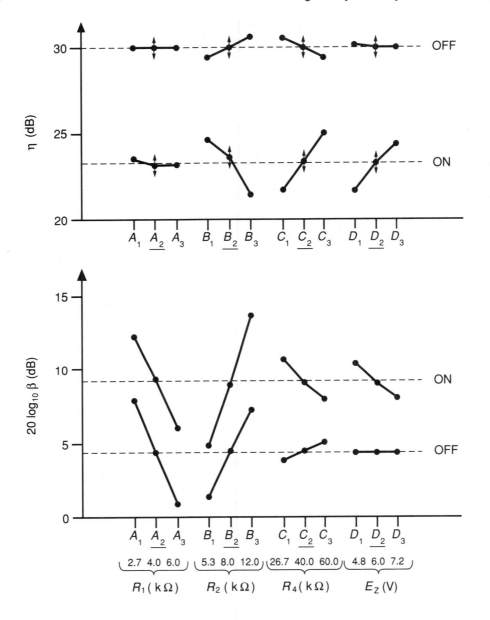

Figure 9.6 Plots of factor effects. Underscore indicates starting level. Two-standard-deviation confidence limits are also shown for the starting level. Estimated confidence limits for $20 \log_{10} \beta$ are too small to show.

Thus, the optimum settings of the control factors suggested by Figure 9.6(a) are $R_1 = 4.0 \ k\Omega$, $R_2 = 5.336 \ k\Omega$, $R_4 = 60.0 \ k\Omega$, and $E_z = 7.2 \ V$. The verification experiment under the optimum conditions gave $\eta = 26.43 \ dB$, compared to 23.84 dB under the starting design. Similarly, under the optimum conditions we obtained $\eta' = 29.10$ dB, compared to 29.94 dB under the starting design. Note that the increase in η is much larger than the reduction in η'.

From Figure 9.6(b), it is clear that both R_1 and R_2 have a large effect on β and β'. The control factors R_4 and E_z have a somewhat less effect on β and β'. Since R_1 has no effect on the S/N ratios, it is an ideal choice for adjusting the slopes β and β'. This ability to adjust β and β' gives the needed freedom to: (1) match the values of R_{T-ON} and R_{T-OFF} with the chosen thermistor for serving the desired temperature range, and (2) obtain the desired hysteresis. As discussed earlier, the needed separation (hysteresis) between R_{T-ON} and R_{T-OFF} is determined by the thermal analysis of the heating system, which is not discussed here.

9.5 ITERATIVE OPTIMIZATION

The preceding section showed one cycle of Robust Design. It is clear from Figure 9.6 that the potential exists for further improvement. By taking the optimum point as a starting design, one can repeat the optimization procedure to achieve this potential—that was indeed done for the temperature control circuit. For each iteration, we took the middle level for each control factor as the optimum level from the previous iteration, and then took levels 1 and 3 to have the same relationship with the level 2 for that factor as in the first iteration. However, during these iterations, we did not let the value of E_z exceed 7.2 V so that adequate separation between E_z and E_0 obtained through three iterations are shown in Table 9.7. Of course, some additional improvement is possible, but by the third iteration, the rate of improvement has clearly slowed down, so one need not proceed further.

The foregoing discussion points to the potential of using orthogonal arrays to optimize nonlinear functions iteratively. Although we would expect this to be a topic of active research, the following advantages of using orthogonal arrays over many commonly used nonlinear programming methods are apparent:

- No derivatives have to be computed

- Hessian does not have to be computed

- Algorithm is less sensitive to starting conditions

- Large number of variables can be handled easily

- Combinations of continuous and discrete variables can be handled easily

While most standard nonlinear programming methods are based on the first- and second-order derivatives of the objective function at a point, the orthogonal array method looks at a wide region in each iteration. That is, while the nonlinear programming methods constitute point approaches, the orthogonal array method is a region approach. Experience in using the orthogonal array method with a variety of problems indicates that, because of the region approach, it works particularly well in the early stages, that is, when the starting point is far from the optimum. Once we get near the optimum point, some of the standard nonlinear programming methods, such as the Newton-Ralphson method, work very well. Thus one may wish to use the orthogonal array method in the beginning and then switch to a standard nonlinear programming method.

TABLE 9.7 ITERATIVE OPTIMIZATION

Iteration Number	η (dB)	η' (dB)	$\eta + \eta'$ (dB)
Starting condition	23.84	29.94	53.78
Iteration 1	26.43	29.10	55.53
Iteration 2	27.30	28.70	56.00
Iteration 3	27.77	28.51	56.28

9.6 SUMMARY

- Dynamic systems are those in which we want the system's response to follow the levels of the signal factor in a prescribed manner. The changing nature of the levels of the signal factor and the response make the design of a dynamic system more complicated than designing a static system. Nevertheless, the eight steps of Robust Design (described in Chapter 4) still apply.

- A temperature controller is a feedback system and can be divided into three main modules: (1) temperature sensor, (2) temperature control circuit, and (3) a heating (or cooling) element. For designing a robust temperature controller, the three modules must be made robust separately and then integrated together.

- The temperature control circuit is a doubly dynamic system. First, for a particular target temperature of the bath, the circuit must turn a heater ON or OFF at specific threshold temperature values. Second, the target temperature may be changed by the user.

- Four circuit parameters (R_1, R_2, R_4, and E_z) were selected as control factors. The resistance R_3 was chosen as the signal factor. The tolerances in the control factors were the noise factors.

- The threshold resistance, R_{T-ON}, at which the heater turns ON and the threshold resistance, R_{T-OFF}, at which the heater turns OFF were selected as the quality characteristics. The variation of R_{T-ON} and R_{T-OFF} as a function of R_3 formed two C-C type dynamic problems.

- To evaluate the S/N ratio (η) for the ON function, a compound noise factor, CN was formed. Three levels were chosen for the signal factor and the compound noise factor. R_{T-ON} was computed at the resulting nine combinations of the signal and noise factor levels and the S/N ratio for the ON function was then computed. (An orthogonal array can be used for computing the S/N ratio when engineering judgment dictates that multiple noise factors be used). The S/N ratio for the OFF function (η') was evaluated in the same manner.

- The L_{18} array was used as the control orthogonal array. Through one cycle of Robust Design, the sum $\eta + \eta'$ was improved by 1.75 dB. Iterating the Robust Design cycle three times led to 2.50 dB improvement in $\eta + \eta'$.

- Orthogonal arrays can be used to optimize iteratively a nonlinear function. They provide a region approach and perform especially well when the starting point is far from the optimum and when some of the parameters are discrete.

Chapter 10

TUNING COMPUTER SYSTEMS FOR HIGH PERFORMANCE

A computer manufacturer invariably specifies the "best" operating conditions for a computer. Why then is tuning a computer necessary? The answer lies in the fact that manufacturers specify conditions based on assumptions about applications and loads. However, actual applications and load conditions might be different or might change over a period of time, which can lead to inferior computer performance. Assuming a system has tunable parameters, two options are available to improve performance: (1) buying more hardware or (2) finding optimum settings of tunable parameters. One should always consider the option of improving the performance of a computer system through better administration before investing in additional hardware. In fact, this option should be considered for any system that has tunable parameters that can be set by the system administrator. Determining the best settings of tunable parameters by the prevailing trial-and-error method may, however, prove to be excessively time consuming and expensive. Robust Design provides a systematic and cost efficient method of experimenting with a large number of tunable parameters, thus yielding greater improvements in performance quicker.

This chapter presents a case study to illustrate the use of the Robust Design method in tuning computer performance. A few details have been modified for pedagogic purposes. The case study was performed by T. W. Pao, C. S. Sherrerd, and M. S. Phadke [P1] who are considered to be the first to conduct such a study to optimize a hardware-software system using the Robust Design method.

There are nine sections in this chapter:

- Section 10.1 describes the problem formulation of the case study (Step 1 of the Robust Design steps described in Chapter 4).

- Section 10.2 discusses the noise factors and testing conditions (Step 2).

- Section 10.3 describes the quality characteristic and the signal-to-noise (S/N) ratio (Step 3).

- Section 10.4 discusses control factors and their alternate levels (Step 4).

- Section 10.5 describes the design of the matrix experiment and the experimental procedure used by the research team (Steps 5 and 6).

- Section 10.6 gives the data analysis and verification experiments (Steps 7 and 8).

- Section 10.7 describes the standardized S/N ratio that is useful in compensating for variation in load during the experiment.

- Section 10.8 discusses some related applications.

- Section 10.9 summarizes the important points of this chapter.

10.1 PROBLEM FORMULATION

The case study concerns the performance of a VAX* 11-780 machine running the UNIX operating system, Release 5.0. The machine had 48 user terminal ports, two remote job entry links, four megabytes of memory, and five disk drives. The average number of users logged on at a time was between 20 and 30.

Before the start of the project, the users' perceptions were that system performance was very poor, especially in the afternoon. For an objective measurement of the response time, the experimenters used two specific, representative commands called *standard* and *trivial*. The standard command consisted of creating, editing, and removing a file; the trivial command was the UNIX system *date* command, which does not involve input/output (I/O). Response times were measured by submitting these commands *via* the UNIX system *crontab* facility and clocking the time taken for the computer to respond using the UNIX system *timex* command, both of which are automatic system processes.

For the particular users of this machine, the response time for the standard and trivial commands could be considered representative of the response time for other

---

* VAX is a registered trademark of Digital Equipment Corporation.

various commands for that computer. In some other computer installations, response time for the compilation of a C program or the time taken for the *troff* command (a text processing command) may be more representative.

Figure 10.1(a) – (b) shows the variation of the response times as functions of time of day for the standard and trivial commands. Note that at the start of the study, the average response time increased as the afternoon progressed (see the curves marked "Initial" in the figure). The increase in response time correlated well with the increase in the work load during the afternoon. The objective in the experiment was to make the response time uniformly small throughout the day, even when the load increased as usual.

There are two broad approaches for optimizing a complex system such as a computer: (1) micro-modeling and (2) macro-modeling. They are explained next.

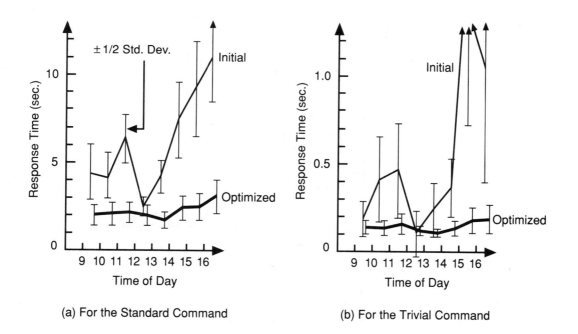

(a) For the Standard Command   (b) For the Trivial Command

Figure 10.1   Comparison of response times.

Micro-Modeling

As the name suggests, micro-modeling is based on an in-depth understanding of the system. It begins by developing a mathematical model of the system, which, in this case, would be of the complex internal queuing of the UNIX operating system. When systems are complex, as in the case study, we must make assumptions that simplify the operation, as well as put forth considerable effort to develop the model. Furthermore, the more simplifying we do, the less realistic the model will be, and, hence, the less adequate it will be for precise optimization. But once an adequate model is constructed, a number of well-known optimization methods, including Robust Design, can be used to find the best system configuration.

Macro-Modeling

In macro-modeling, we bypass the step of building a mathematical model of the system. Our concern is primarily with obtaining the optimum system configuration, not with obtaining a detailed understanding of the system itself. As such, macro-modeling gives faster and more efficient results. It gives the specific information needed for optimization with a minimum expenditure of experimental resources.

The UNIX system is viewed as a "black box," as illustrated in Figure 10.2. The parameters that influence the response time are identified and divided into two classes: noise factors and control factors. The best settings of the control factors are determined through experiments. Thus, the Robust Design method lends itself well for optimization through the macro-modeling approach.

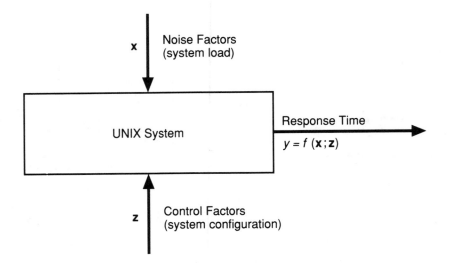

Figure 10.2   Block diagram for the UNIX system.

10.2 NOISE FACTORS AND TESTING CONDITIONS

Load variation during use of the machine, from day-to-day and as a function of the time of day, constitutes the main noise factor for the computer system under study.

The number of users logged on, central processor unit (CPU) demand, I/O demand, and memory demand are some of the more important load measures. Temperature and humidity variations in the computer room, as well as fluctuations in the power supply voltage, are also noise factors but are normally of minor consequence.

The case study was conducted live on the computer. As a result, the normal variation of load during the day provided the various testing conditions for evaluating the S/N ratio.

At the beginning of the study, the researchers examined the operating logs for the previous few weeks to evaluate the day-to-day fluctuation in response time and load. The examination revealed that the response time and the load were roughly similar for all five weekdays. This meant that Mondays could be treated the same as Tuesdays, etc. If the five days of the week had turned out to be markedly different from each other, then those differences would have had to be taken into account in planning the experiment.

10.3 QUALITY CHARACTERISTIC AND S/N RATIO

Let us first consider the standard command used in the study. Suppose it takes t_0 seconds to execute that command under the best circumstances—that is, when the load is zero, t_0 is the minimum possible time for the command. Then, it becomes obvious that the actual response time for the standard command minus t_0 is a quality characteristic that is always nonnegative and has a target value of zero—that is, the actual response time minus t_0 belongs to the smaller-the-better type problems. In the case study, the various measurements of response time showed that t_0 was much smaller than the observed response time. Hence, t_0 was ignored and the measured response time was treated as a smaller-the-better type characteristic. The corresponding S/N ratio to be maximized is

$$\eta = -10 \log_{10} \begin{bmatrix} \text{mean square response time} \\ \text{for the standard command} \end{bmatrix}. \qquad (10.1)$$

Referring to Figure 10.1, it is clear that at the beginning the standard deviation of the response time was large, so much so that it is shown by bars of length $\pm 1/2$ standard deviation, as opposed to the standard practice of showing ± 2 standard deviations. From the quadratic loss function considerations, reducing both the mean and variance is important. It is clear that the S/N ratio in Equation (10.1) accomplishes this goal because mean square response time is equal to sum of the square of the mean and the variance.

For the response time for the trivial command, the same formulation was used. That is, the S/N ratio was defined as follows:

$$\eta' = -10 \log_{10} \left[\begin{array}{c} \text{mean square response time} \\ \text{for the trivial command} \end{array} \right]. \qquad (10.2)$$

10.4 CONTROL FACTORS AND THEIR ALTERNATE LEVELS

The UNIX operating system provides a number of tunable parameters, some of which relate to the hardware and others to the software. Through discussions with a group of system administrators and computer scientists, the experiment team decided to include the eight control factors listed in Table 10.1 for the tuning study. Among them, factors A, C, and F are hardware related, and the others are software related. A description of these parameters and their alternate levels is given next. The discussion about the selection of levels is particularly noteworthy because it reveals some of the practical difficulties faced in planning and carrying out Robust Design experiments.

TABLE 10.1 CONTROL FACTORS AND LEVELS

	Levels*		
Factor	**1**	**2**	**3**
A. Disk drives (RMO5 & RPO6)	4&1	4&2	4&2
B. File distribution	a	b	c
C. Memory size (MB)	4	3	3.5
D. System buffers	1/5	1/4	1/3
E. Sticky bits	0	3	8
F. KMCs used	2	0	
G. INODE table entries	400	500	600
H. Other system tables	a	b	c

* The starting levels are indicated by an underscore.

The number and type of *disk drives* (factor A) is an important parameter that determines the I/O access time. At the start, there were four RMO5 disks and one RPO6 disk. The experimenters wanted to see the effect of adding one more RPO6 disk (level A_2), as well as the effect of adding one RPO7 disk and a faster memory controller (level A_3). However, the RPO7 disk did not arrive in time for the experiments. So, level A_3 was defined to be the same as level A_2 for factor A. The next section discusses the care taken in planning the matrix experiment, which allowed the experimenters to change the plan in the middle of the experiments.

The *file system distributions* (factor B) a, b and c refer to three specific algorithms used for distributing the user and system files among the disk drives. Obviously, the actual distribution depends on the number of disk drives used in a particular system configuration. Since the internal entropy (a measure of the lack of order in storing the files) could have a significant impact on response time, the team took care to preserve the internal entropy while changing from one file system to another during the experiments.

One system administrator suggested increasing the *memory size* (factor C) to improve the response time. However, another expert opinion was given that stated additional memory would not improve the response for the particular computer system being studied. Therefore, the team decided not to purchase more memory until they were reasonably sure its cost would be justified. They took level C_1 as the existing memory size, namely 4 MB, and disabled some of the existing memory to form levels C_2 and C_3. They decided to purchase more memory only if the experimental data showed that disabling a part of the memory leads to a significant reduction in performance.

Total memory is divided into two parts: *system buffers* (factor D) and user memory. The system buffers are used by the operating system to store recently used data in the hope that the data might be needed again soon. Increasing the size of the system buffers improves the probability (technically called *hit ratio*) of finding the needed data in the memory. This can contribute to improved performance, but it also reduces the memory available to the users, which can lead to progressively worse system performance. Thus, the optimum size of system buffers can depend on the particular load pattern. We refer to the size of the system buffers as a fraction of the total memory size. Thus, the levels of the system buffers are sliding with respect to the memory size.

Sticky bit (factor E) is a way of telling an operating system to treat a command in a special way. When the sticky bit for a command such as *rm* or *ed* is set, the executable module for that command is copied contiguously in the swap area of a disk during system initialization. Every time that command is needed but not found in the memory, it is brought back from the swap area expeditiously in a single operation. However, if the sticky bit is not set and the command is not found in the memory, it must be brought back block by block from the file system. This adds to the execution time.

In this case study, factor E specifies how many and which commands had their sticky bits set. For level E_1, no command had its sticky bit set. For level E_2, the three commands that had their sticky bits set were *sh, ksh* and *rm*. These were the three most frequently used commands during the month before the case study according to a 5-day accounting command summary report. For the level E_3, the eight commands that had their sticky bits set were the three commands mentioned above, plus the next five most commonly used commands, namely, the commands *ls, cp, expr, chmod,* and *sadc* (a local library command).

KMCs (factor F) are special devices used to assist the main CPU in handling the terminal and remote job entry traffic. They attempt to reduce the number of interrupts faced by the main CPU. In this case study, only the KMCs used for terminal traffic were changed. Those used for the remote job entry links were left alone. For level F_1, two KMCs were used to handle the terminal traffic, whereas for level F_2 the two KMCs were disabled.

The number of entries in the *INODE table* (factor G) determines the number of user files that can be handled simultaneously by the system. The three levels for the factor G are 400, 500, and 600. The three levels of the eighth factor, namely, the *other system tables* (factor H), are coded as *a*, *b*, and *c*.

Note that the software factors (B, D, E, G, and H) can affect only response time. However, the three hardware factors (A, C, and F) can affect both the response time and the computer cost. Therefore, this optimization problem is not a pure parameter design problem, but, rather, a hybrid of parameter design and tolerance design.

10.5 DESIGN OF THE MATRIX EXPERIMENT AND THE EXPERIMENTAL PROCEDURE

This case study has seven 3-level factors and one 2-level factor. There are $7 \times (3-1) + 1 \times (2-1) + 1 = 16$ degrees of freedom associated with these factors. The orthogonal array L_{18} is just right for this project because it has seven 3-level columns and one 2-level column to match the needs of the matrix experiment. The L_{18} array and the assignment of columns to factors are shown in Table 10.2. Aside from assigning the 2-level factor to the 2-level column, there is really no other reason for assigning a particular factor to a particular column. The factors were assigned to the columns in the order in which they were listed at the time the experiment was planned. Some aspects of the assignment of factors to columns are discussed next.

The experiment team found that changing the level of disk drives (factor A) was the most difficult among all the factors because it required an outside technician and took three to four hours. Consequently, in conducting these experiments, the team first conducted all experiments with level A_1 of the disk drives, then those with level A_2 of the disk drives, and finally those with level A_3 of the disk drives. The experiments with level A_3 of the disk drives were kept for last to allow time for the RPO7 disk to arrive. However, because the RPO7 disk did not arrive in time, the experimenters redefined level A_3 of the disk drives to be the same as level A_2 and continued with the rest of the plan. According to the dummy level technique discussed in Chapter 7, this redefinition of level does not destroy the orthogonality of the matrix experiment. This arrangement, however, gives 12 experiments with level A_2 of the disk drives; hence, more accurate information about that level is obtained when compared to level A_1. This is exactly what we should look for because level A_2 is the new level about which we have less prior information.

TABLE 10.2 L_{18} ORTHOGONAL ARRAY AND FACTOR ASSIGNMENT

Expt. No.	Column Number and Factor Assignment							
	1 F	2 B	3 C	4 D	5 E	6 A	7 G	8 H
1	1	1	1	1	1	1	1	1
2	1	1	2	2	2	2	2	2
3	1	1	3	3	3	3	3	3
4	1	2	1	1	2	2	3	3
5	1	2	2	2	3	3	1	1
6	1	2	3	3	1	1	2	2
7	1	3	1	2	1	3	2	3
8	1	3	2	3	2	1	3	1
9	1	3	3	1	3	2	1	2
10	2	1	1	3	3	2	2	1
11	2	1	2	1	1	3	3	2
12	2	1	3	2	2	1	1	3
13	2	2	1	2	3	1	3	2
14	2	2	2	3	1	2	1	3
15	2	2	3	1	2	3	2	1
16	2	3	1	3	2	3	1	2
17	2	3	2	1	3	1	2	3
18	2	3	3	2	1	2	3	1

File distribution (factor B) was the second most difficult factor to change. Therefore, among the six experiments with level A_1 of the disk drives, the experiment team first conducted the two experiments with level B_1, then those with level B_2, and finally those with level B_3. The same pattern was repeated for level A_2 of the disk drives and then level A_3 of the disk drives. The examination of the L_{18} array given in Table 10.2 indicates that some of the bookkeeping of the experimental conditions could have been simplified if the team had assigned factor A to Column 2 and factor B to

Column 3. However, the inconvenience of the particular column assignment was unimportant.

For each experiment corresponding to each row of the L_{18} array, the team ran the system for two days and collected data on standard response time and trivial response time once every 10 minutes from 9:00 a.m. to 5:00 p.m. While measuring the response times, they made sure that the UNIX system *cron* facility did not schedule some routine data collection operations at the same exact moments because this would affect the measurement of response times.

Running experiments on a live system can invite a number of practical problems. Therefore, the first thing the experimenters did was to seek the cooperation of users. One problem of great concern to the users was that for a particular combination of control factor settings, the system performance could become bad enough to cause a major disruption of their activities. To avoid this, the system administrator was instructed to make note of such an event and go back to the pre-experiment settings of the various factors. This would minimize the inconvenience to the users. Fortunately, such an event did not occur, but had it happened, the experiment team would still have been able to analyze the data and determine optimum conditions using the accumulation analysis method described in Chapter 5 (see also Taguchi [T1], and Taguchi and Wu [T7]).

Under the best circumstances, the team could finish two experiments per week. For 18 experiments, it would then take nine weeks. However, during the experiments, a snowstorm arrived and the Easter holiday was observed, both events causing the system load to drop to an exceptionally low level. Those days were eliminated from the data, and the team repeated the corresponding combinations of control factor settings to have data for every row of the matrix experiment.

10.6 DATA ANALYSIS AND VERIFICATION EXPERIMENTS

From the 96 measurements of standard response time for each experiment, the team computed the mean response time and the S/N ratio. The results are shown in Table 10.3. Similar computations were made for the trivial response time, but they are not shown here. The effects of the various factors on the S/N ratio for the standard response time are shown, along with the corresponding ANOVA in Table 10.4. The factor effects are plotted in Figure 10.3. Note that the levels C_1, C_2, and C_3 of memory size are 4.0 MB, 3.0 MB, and 3.5 MB, respectively, which are not in a monotonic order. While plotting the data in Figure 10.3, the experimenters considered the correct order.

It is apparent from Table 10.4 that the factor effects are rather small, especially when compared to the error variance. The problem of getting large error variance is more likely with live experiments, such as this computer system optimization experiment, because the different rows of the matrix experiment are apt to see quite different noise conditions, that is, quite different load conditions. Also, while running live

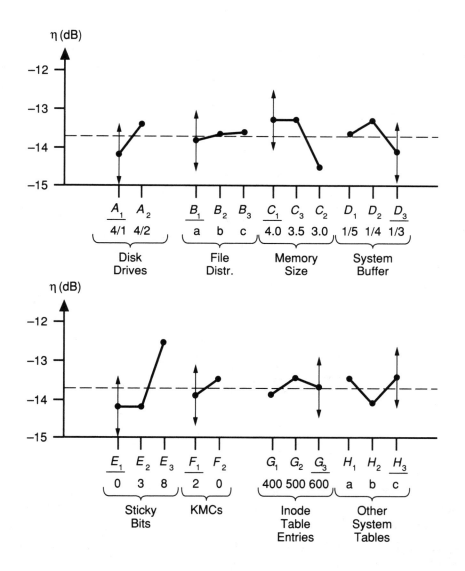

Figure 10.3   Factor effects for S/N ratio for standard response. Underscore indicates starting level. One-standard-deviation limits are also shown.

experiments, the tendency is to choose the levels of control factors that are not far apart. However, we can still make valuable conclusions about optimum settings of control factors and then see if the improvements observed during the verification experiment are significant or not.

TABLE 10.3 DATA SUMMARY FOR STANDARD RESPONSE TIME

Expt. No.	Mean (sec)	η (dB)
1	4.65	−14.66
2	5.28	−16.37
3	3.06	−10.49
4	4.53	−14.85
5	3.26	−10.94
6	4.55	−14.96
7	3.37	−11.77
8	5.62	−16.72
9	4.87	−14.67
10	4.13	−13.52
11	4.08	−13.79
12	4.45	−14.19
13	3.81	−12.89
14	5.87	−16.75
15	3.42	−11.65
16	3.66	−12.23
17	3.92	−12.81
18	4.42	−13.71

The following observations can be made from the plots in Figure 10.3 and Table 10.4 (note that these conclusions are valid only for the particular load characteristics of the computer being tuned):

1. Going from not setting any sticky bits to setting sticky bits on the three most used commands does not improve the response time. This is probably because these three commands tend to stay in the memory as a result of their very frequent use, regardless of setting sticky bits. However, when sticky bits are set on the five next most used commands, the response time improves by 1.69 dB. This suggests that we should set sticky bits on the eight commands, and in future experiments, we should consider even more commands for setting sticky bits.

2. KMCs do not help in improving response time for this type of computer environment. Therefore, they may be dropped as far as terminal handling is concerned, thus reducing the cost of the hardware.

TABLE 10.4 ANALYSIS OF S/N RATIOS FOR STANDARD RESPONSE*

Factor	Average η by Factor Level			Degrees of Freedom	Sum of Squares	Mean Square	F
	1	**2**	**3**				
A. Disk drives	−14.37	−13.40		1	3.76	3.76	1
B. File distribution	−13.84	−13.67	−13.65	2	0.12†	0.06	
C. Memory size	−13.32	−14.56	−13.28	2	6.40	3.20	
D. System buffers	−13.74	−13.31	−14.11	2	1.92	0.97	
E. Sticky bits	−14.27	−14.34	−12.55	2	12.27	6.14	2
F. KMCs used	−13.94	−13.50		1	0.84†	0.42	
G. INODE table entries	−13.91	−13.51	−13.74	2	0.47†	0.24	
H. Other system tables	−13.53	−14.15	−13.48	2	1.68	1.34	
Error				3	32.60	10.87	
Total				17	60.06	3.53	
(Error)				(9)	(34.03)	(3.78)	

* Overall mean η = −13.72 dB; underscore indicates starting conditions.
† Indicates the sum of squares that were added to form the pooled error sum of squares
  shown in parentheses.

3. Adding one more disk drive leads to better response time. Perhaps even more
    disks should be considered for improving the response time. Of course, this
    would mean more cost, so proper trade-offs would have to be made.

4. The S/N ratio is virtually the same for 4 MB and 3.5 MB memory. It is
    significantly lower for 3 MB memory. Thus, 4 MB seems to be an optimum
    value—that is, buying more memory would probably not help much in improv-
    ing response time.

5. There is some potential advantage (0.8 dB) in changing the fraction of system
    buffers from 1/3 to 1/4.

6. The effects of the remaining three control factors are very small and there is no
    advantage in changing their levels.

Optimum Conditions and Verification Experiment

The optimum system configuration inferred from the data analysis above is shown in Table 10.5 along with the starting configuration. Changes were recommended in the settings of sticky bits, disk drives, and system buffers because they lead to faster response. KMCs were dropped because they did not help improve response, and dropping them meant saving hardware. The prediction of the S/N ratio for the standard response time under the starting and optimum conditions is also shown in Table 10.5. Note that the contributions of the factors, whose sum of squares were among the smallest and were pooled, are ignored in predicting the S/N ratio. Thus, the S/N ratio predicted by the data analysis under the starting condition is -14.67 dB, and under the optimum conditions it is -11.22 dB. The corresponding, predicted rms response times under the starting and optimum conditions are 5.41 seconds and 3.63 seconds, respectively.

TABLE 10.5 OPTIMUM SETTINGS AND PREDICTION OF STANDARD RESPONSE TIME

Factor	Starting Condition		Optimum Condition	
	Setting	Contribution† (dB)	Setting	Contribution† (dB)
A. Disk drives*	A_1	-0.65	A_2	0.32
B. File distribution	B_1	–	B_1	–
C. Memory size	C_1	0.40	C_1	0.40
D. System buffers*	D_3	-0.39	D_2	0.41
E. Sticky bits*	E_1	-0.55	E_3	1.17
F. KMCs used*	F_1	–	F_2	–
G. INODE table entries	G_3	–	G_3	–
H. Other system tables	H_3	0.24	H_3	0.24
Overall mean		-13.72		-13.72
Total		-14.67		-11.22

* Indicates the factors whose levels are changed from the starting to the optimum conditions.
† By contribution we mean the deviation from the overall mean caused by the particular factor level.

As noted earlier, here the error variance is large. We would also expect the variance of the prediction error to be large. The variance of the prediction error can be computed by the procedure given in Chapter 3 [see Equation (3.14)]. The equivalent sample size for the starting condition, n_e, is given by

$$\frac{1}{n_e} = \frac{1}{n} + \left[\frac{1}{n_{A_1}} - \frac{1}{n} \right] + \left[\frac{1}{n_{C_1}} - \frac{1}{n} \right] + \left[\frac{1}{n_{D_3}} - \frac{1}{n} \right]$$

$$+ \left[\frac{1}{n_{E_1}} - \frac{1}{n} \right] + \left[\frac{1}{n_{H_1}} - \frac{1}{n} \right]$$

$$= \frac{1}{18} + \left[\frac{1}{6} - \frac{1}{18} \right] + \left[\frac{1}{6} - \frac{1}{18} \right] + \left[\frac{1}{6} - \frac{1}{18} \right]$$

$$+ \left[\frac{1}{6} - \frac{1}{18} \right] + \left[\frac{1}{6} - \frac{1}{18} \right]$$

$$= 0.61 .$$

Thus, $n_e = 1.64$. Correspondingly, the two-standard-deviation confidence limits for the predicted S/N ratio under the starting conditions are $-14.67 \pm 2 \sqrt{3.78/1.64}$, which simplify to -14.67 ± 3.04 dB. Similarly, the two-standard-deviation confidence limits for the predicted S/N ratio under the optimum conditions are $-11.22 \pm 2\sqrt{3.78/1.89}$, which simplify to -11.22 ± 2.83 dB. Note that there is a slight difference between the widths of the two confidence intervals. This is because there are 12 experiments with level A_2 and only six with A_1.

Subsequently, the optimum configuration was implemented. The average response times for the standard and trivial commands for over two weeks of operation under the optimum conditions are plotted in Figure 10.1(a) – (b). Comparing the response time under the starting (initial) configuration with that under the optimum configuration, we see that under the optimum conditions the response time is small, even in the afternoon when the load is high. In fact, the response time is uniformly low throughout the day.

The data from the confirmation experiment are summarized in Table 10.6. For standard response time we see that the mean response time was reduced from 6.15 sec. to 2.37 sec., which amounts to a 61-percent improvement. Similar improvement was seen in the rms response time. On the S/N ratio scale, the improvement was 8.39 dB. Similarly, for the trivial response time, the improvement was seen to be between 70 percent and 80 percent of the mean value, or 13.64 dB, as indicated in Table 10.6.

TABLE 10.6 RESULTS OF VERIFICATION EXPERIMENT

Measure	Standard Response Time			Trivial Response Time		
	Starting Levels	Optimum Levels	Improvement	Starting Levels	Optimum Levels	Improvement
Mean (sec)	6.15	2.37	61%	0.521	0.148	71%
rms (sec)	7.59	2.98	61%	0.962	0.200	79%
η (dB)	−17.60	−9.21	8.39 dB	+0.34	+13.98	13.64 dB
Standardized η (dB)	−15.88	−8.64	7.24 dB	+3.26	+15.71	12.45 dB

Note that the observed S/N ratios for standard response time under the starting and optimum conditions are within their respective two-standard-deviation confidence limits. However, they are rather close to the limits. Also, the observed improvement (8.39 dB) in the S/N ratio is quite large compared to the improvement predicted by the data (3.47 dB). However, the difference is well within the confidence limits. Also, observe that the S/N ratio under the optimum conditions is better than the best among the 18 experiments.

Thus, here we achieved a 60- to 70-percent improvement in response time by improved system administration. Following this experiment, two similar experiments were performed by Klingler and Nazaret [K5], who took extra care to ensure that published UNIX system tuning guides were used to establish the starting conditions. Their experiments still led to a 20- to 40-percent improvement in response time. One extra factor they considered was the use of PDQs, which are special auxiliary processors for handling text processing jobs (troff). For their systems, it turned out that the use of PDQs could hurt the response time rather than help.

10.7 STANDARDIZED S/N RATIO

When running Robust Design experiments with live systems, we face the methodological problem that the noise conditions, which are the load conditions for our computer system optimization, are not the same for every row of the control orthogonal array. This can lead to inaccuracies in the conclusions. One way to minimize the impact of changing noise conditions is to construct a standardized S/N ratio, which we describe next.

As noted earlier, some of the more important load measures for the computer system optimization experiment are: number of users, CPU demand, I/O demand, and memory demand. After studying the load pattern over the days when the case study was conducted, we can define low and high levels for each of these load measures, as shown in Figure 10.4. These levels should be defined so that for every experiment we have a reasonable number of observations at each level. The 16 different possible combinations of the levels of these four load measures are listed in Table 10.7 and are nothing more than 16 different noise conditions.

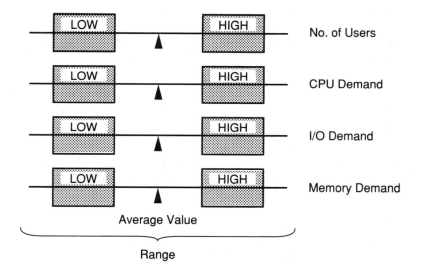

Figure 10.4   Selection of levels for load measures.

In a live experiment, the number of observations for each noise condition can change from experiment to experiment. For instance, one day the load might be heavy while another day it might be light. Although we cannot dictate the load condition for the different experiments, we can observe the *offered load*. The impact of load variation can be minimized as follows: We first compute the average response time for each experiment in each of the 16 load conditions. We then treat these 16 averages as raw data to compute the S/N ratio for each experiment. The S/N ratio computed in this manner is called standardized S/N ratio because it effectively standardizes the load conditions for each experiment. The system can then be optimized using this standardized S/N ratio.

Note that for the standardized S/N ratio to work, we must have good definitions of noise factors and ways of measuring them. Also, in each experiment every noise condition must occur at least once to be able to compute the average. In practice,

TABLE 10.7 COMPUTATION OF STANDARDIZED S/N RATIO*

Expt. No.	No. of Users	CPU Demand	I/O Demand	Memory Demand	Average Response Time for Experiment i
1	1	1	1	1	$y_{i\,1}$
2	1	1	1	2	$y_{i\,2}$
3	1	1	2	1	$y_{i\,3}$
4	1	1	2	2	$y_{i\,4}$
5	1	2	1	1	$y_{i\,5}$
6	1	2	1	2	$y_{i\,6}$
7	1	2	2	1	$y_{i\,7}$
8	1	2	2	2	$y_{i\,8}$
9	2	1	1	1	$y_{i\,9}$
10	2	1	1	2	$y_{i\,10}$
11	2	1	2	1	$y_{i\,11}$
12	2	1	2	2	$y_{i\,12}$
13	2	2	1	1	$y_{i\,13}$
14	2	2	1	2	$y_{i\,14}$
15	2	2	2	1	$y_{i\,15}$
16	2	2	2	2	$y_{i\,16}$

* Standardized S/N ratio for experiment i,

$$\eta_i = -10 \log_{10} \left[\frac{1}{16} \sum_{j=1}^{16} y_{ij}^2 \right]$$

however, if one or two conditions are missing, we may compute the S/N ratio with the available noise conditions without much harm.

In the computer system optimization case study discussed in this chapter, the experimenters used the concept of standardized S/N ratio to obtain better comparison of the starting and optimum conditions. Some researchers expressed a concern that part of the improvement observed in this case study might have been due to changed load conditions over the period of three months that it took the team to conduct the matrix experiment. Accordingly, the experimenters computed the standardized S/N ratio for the two conditions and the results are shown in Table 10.6 along with the other results of the verification experiment. Since the improvement in the standardized S/N ratio is quite close to that in the original S/N ratio, the experiment team concluded that the load change had a minimal impact on the improvement.

10.8 RELATED APPLICATIONS

Computer System Parameter Space Mapping

In practice, running live experiments to optimize every computer installation is obviously not practicable. Instead, more benefits can be achieved by performing off-line matrix experiments to optimize for several different load conditions. Such load conditions can be simulated using facilities like the UNIX system benchmarking facility. The information gained from these off-line experiments can be used to map the operating system parameter space. This catalog of experimental results can then be used to improve the performance of different machines.

Optimizing Operations Support Systems

Large systems, such as the traffic in a telecommunications network, are often managed by software systems that are commonly known as *operations support systems*. The load offered to a telephone network varies widely from weekday to weekend, from morning to evening, from working days to holidays, etc. Managing the network includes defining strategies for providing the needed transmission and switching facilities, routing a call end to end on the network, and many other tasks. Steve Eick and Madhav Phadke applied the Robust Design method in AT&T to improve the strategy for adapting the routing algorithm to handle unusually high demands. Studying eight parameters of the strategy using the L_{18} orthogonal array, they were able to increase the excess load carried during unusually high demands by as much as 70 percent. This not only improved the revenue, but also improved network performance by reducing the number of blocked calls. These experiments were done on a simulator because of the high risks of the live experiments.

International telephone traffic is carried over several different types of transmission facilities. Seshadri [S4] used the Robust Design method to determine the preferred facility for transmitting facsimile data by conducting live experiments on the network.

The success in improving the telephone network traffic management suggests that the Robust Design approach could be used successfully to improve other networks, for example, air traffic management. In fact, it is reported that the method has been used in Japan to optimize runway and air-space usage at major airports.

10.9 SUMMARY

- The performance of complex systems (such as a computer), which have tunable parameters, can often be improved through better system administration. Tuning is necessary because the parameter settings specified by the system manufacturer pertain to specific assumptions regarding the applications and load. However, the actual applications and load could be different from those envisioned by the manufacturer.

- There are two systematic approaches for optimizing a complex system: (1) micro-modeling and (2) macro-modeling. The macro-modeling approach can utilize the Robust Design method to achieve more rapid and efficient results.

- Load variation during use of the computer, from day to day and as a function of the time of day, constitutes the main noise factor for the computer system. The number of users logged on, CPU demand, I/O demand, and memory demand are some of the more important load measures.

- The response time for the standard (or the trivial) command minus the minimum possible time for that command was the quality characteristic for the case study. It is a smaller-the-better type characteristic. The minimum possible response time was ignored in the analysis as it was very small compared to the average response time.

- Eight control factors were chosen for the case study: disk drives (A), file distribution (B), memory size (C), system buffers (D), sticky bits (E), KMCs used (F), INODE table entries (G), and other system tables (H). Factors A, C, and F are hardware related, whereas the others are software related. Factor F had two levels while the others had three levels.

- The L_{18} orthogonal array was used for the matrix experiment. Disk drives was the most difficult factor to change, and file distribution was the next most difficult factor to change. Therefore, the 18 experiments were conducted in an order that minimized changes in these two factors. Also, the experiments were ordered in a way that allowed changing level 3 of disk drives, as anticipated during the planning stage.

- The experiments were conducted on a live system. Each experiment lasted for two days, with eight hours per day. Response time for the standard and the trivial commands was observed once every 10 minutes by using an automatic measurement facility.

- Running Robust Design experiments on live systems involves a methodological problem that the noise conditions (which are the load conditions in the computer response time case study) are not the same for every row of the matrix experiment. Consequently, the estimated error variance can be large compared to the estimated factor effects. Also, the predicted response under the optimum conditions can have large variance. Indeed, this was observed in the present case study.

- Standardized S/N ratios can be used to reduce the adverse effect of changing noise conditions, which is encountered in running Robust Design experiments on live systems.

- Data analysis indicated that levels of four control factors (A, D, E, and F) should be changed and that the levels of the other four factors should be kept the same. It also indicated that the next round of experiments should consider setting sticky bits on more than eight commands.

- The verification experiment showed a 61-percent improvement in both the mean and rms response time for the standard command. This corresponds to an 83-percent reduction in variance of the response time. It also showed a 71-percent improvement in mean response time and a 79-percent improvement in rms response time for the trivial command.

- Computer response time optimization experiments can be conducted off-line using automatic load generation facilities such as the UNIX system benchmarking facility. Off-line experiments are less disruptive for the user community. Orthogonal arrays can be used to determine the load conditions to be simulated.

Chapter 11

RELIABILITY IMPROVEMENT

Increasing the longevity of a product, the time between maintenance of a manufacturing process, or the time between two tool changes are always important considerations in engineering design. These considerations are often lumped under the term *reliability improvement*. There are three fundamental ways of improving the reliability of a product during the design stage: (1) reduce the sensitivity of the product's function to the variation in product parameters, (2) reduce the rate of change of the product parameters, and (3) include redundancy. The first approach is nothing more than the parameter design described earlier in this book. The second approach is analogous to tolerance design and it typically involves more expensive components or manufacturing processes. Thus, this approach should be considered only after sensitivity has been minimized. The third approach is used when the cost of failure of the product is high compared to the cost of providing redundant components or even the whole product.

The most cost-effective approach for reliability improvement is to find appropriate continuous quality characteristics and reduce their sensitivity to all noise factors. Guidelines for selecting quality characteristics related with product reliability were discussed in Chapter 6 in connection with the design of the paper handling system in copying machines. As noted in Chapter 6, determining such quality characteristics is not always easy with existing engineering know-how. *Life tests*, therefore, must be performed to identify settings of control factors that lead to longer product life. This chapter shows the use of the Robust Design methodology in reliability improvement through a case study of router bit life improvement conducted by Dave Chrisman and Madhav Phadke, documented in Reference [P3].

This chapter has nine sections:

- Section 11.1 describes the role of signal-to-noise (S/N) ratios in reliability improvement.

- Section 11.2 describes the routing process and the goal for the case study (Step 1 of the eight Robust Design steps described in Chapter 4).

- Section 11.3 describes the noise factors and quality characteristic (Steps 2 and 3).

- Section 11.4 gives the control factors and their alternate levels (Step 4).

- Section 11.5 describes the construction of the control orthogonal array (Step 5). The requirements for the project were such that the construction of the control orthogonal array was a complicated combinatoric problem.

- Section 11.6 describes the experimental procedure (Step 6).

- Section 11.7 gives the data analysis, selection of optimum conditions, and the verification experiment (Steps 7 and 8).

- Section 11.8 discusses estimation of the factor effects on the survival probability curves.

- Section 11.9 summarizes the important points of this chapter.

11.1 ROLE OF S/N RATIOS IN RELIABILITY IMPROVEMENT

First, let us note the difference between reliability characterization and reliability improvement. *Reliability characterization* refers to building a statistical model for the failure times of the product. Log-normal and Weibull distributions are commonly used for modeling the failure times. Such models are most useful for predicting warranty cost. *Reliability improvement* means changing the product design, including the settings of the control factors, so that time to failure increases.

Invariably, it is expensive to conduct life tests so that an adequate failure-time model can be estimated. Consequently, building adequate failure-time models under various settings of control parameters, as in an orthogonal array experiment, becomes impractical and, hence, is hardly ever done. In fact, it is recommended that conducting life tests should be reserved as far as possible only for a final check on a product. Accelerated life tests are well-suited for this purpose.

For improving a product's reliability, we should find appropriate quality characteristics for the product and minimize its sensitivity to all noise factors. This automatically increases the product's life. The following example clarifies the relationship between the life of a product and sensitivity to noise factors.

Consider an electrical circuit whose output voltage, y, is a critical characteristic. If it deviates too far from the target, the circuit's function fails. Suppose the variation in a resistor, R, plays a key role in the variation of y. Also, suppose, the resistance R is sensitive to environmental temperature and that the resistance increases at a certain rate with aging. During the use of the circuit, the ambient temperature may go too high or too low, or sufficient time may pass leading to a large deviation in R. Consequently, the characteristic y would go outside the limits and the product would fail. Now, if we change the nominal values of appropriate control factors, so that y is much less sensitive to variation in R, then for the same ambient temperatures faced by the circuit and the same rate of change of R due to aging, we would get longer life out of that circuit.

Sensitivity of the voltage y to the noise factors is measured by the S/N ratio. Note that in experiments for improving the S/N ratio, we may use only temperature as the noise factor. Reducing sensitivity to temperature means reducing sensitivity to variation in R and, hence, reducing sensitivity to the aging of R also. Thus, by appropriate choice of testing conditions (noise factor settings) during Robust Design experiments, we can improve the product life as well.

Estimation of Life Using a Benchmark Product

Estimating the life of a newly designed product under customer-usage conditions is always a concern for the product designer. It is particularly important to estimate the life without actually conducting field studies because of the high expense and the long delay in getting results. Benchmark products and the S/N ratio can prove to be very useful in this regard. A benchmark product could be the earlier version of that product with which we have a fair amount of field experience.

Suppose we test the current product and the benchmark product under the same set of testing conditions. That is, we measure the quality characteristic under different levels of the noise factors. For example, in the circuit example above, we may measure the resistance R at the nominal temperature and one or more elevated temperatures. The test may also include holding the circuit at an elevated temperature for a short period of time and then measuring R. By analyzing the test data, suppose we find the S/N ratio for the current product as η_1, and for the benchmark product as η_2. Then, the sensitivity of the quality characteristic for the current product is $\eta_1 - \eta_2$ dB lower than that for the benchmark product—that is, the variance for the current product is smaller by a factor of r, where r is given by

$$r = 10^{\left(\frac{\eta_1 - \eta_2}{10}\right)}.$$

It is often the case that the rate of drift of a product's quality characteristic is proportional to the sensitivity of the quality characteristic to noise factors. Also, the drift in the quality characteristic as a function of time can be approximated reasonably well by the Wiener process. Then, through standard theory of level crossing, we can infer that the average life of the current product would be r times longer than the life of the benchmark product whose average life is known through past experience. Thus, the S/N ratio permits us to estimate the life of a new product in a simple way without conducting expensive and time-consuming life tests.

This section described the role of S/N ratios in reliability improvement. This is a more cost-effective and, hence, preferred way to improve the reliability of a product or a process. However, for a variety of reasons (including lack of adequate engineering know-how about the product) we are forced, in some situations, to conduct life tests to find a way of improving reliability. In the remaining sections of this chapter, we describe a case study of improving the life of router bits by conducting life studies.

11.2 THE ROUTING PROCESS

Typically, printed wiring boards are made in panels of 18×24 in. size. Appropriate size boards, say 8×4 in., are cut from the panels by stamping or by the routing process. A benefit of the routing process is that it gives good dimensional control and smooth edges, thus reducing friction and abrasion during the circuit pack insertion process. When the router bit gets dull, it produces excessive dust which then cakes on the edges of the boards and makes them rough. In such cases, a costly cleaning operation is necessary to smooth the edges. However, changing the router bits frequently is also expensive. In the case study, the objective was to increase the life of the router bits, primarily with regard to the beginning of excessive dust formation.

The routing machine used had four spindles, all of which were synchronized in their rotational speed, horizontal feed (X-Y feed), and vertical feed (in-feed). Each spindle did the routing operation on a separate stack of panels. Typically, two to four panels are stacked together to be cut by each spindle. The cutting process consists of lowering the spindle to an edge of a board, cutting the board all around using the X-Y feed of the spindle, and then lifting the spindle. This is repeated for each board on a panel.

11.3 NOISE FACTORS AND QUALITY CHARACTERISTICS

Some of the important noise factors for the routing process are the out-of-center rotation of the spindle, the variation from one router bit to another, the variation in the material properties within a panel and from panel to panel, and the variation in the speed of the drive motor.

Ideally, we should look for a quality characteristic that is a continuous variable related to the energy transfer in the routing process. Such a variable could be the wear of the cutting edge or the change in the cutting edge geometry. However, these variables are difficult to measure, and the researchers wanted to keep the experiment simple. Therefore, the amount of cut before a bit starts to produce an appreciable amount of dust was used as the quality characteristic. This is the useful life of the bit.

11.4 CONTROL FACTORS AND THEIR LEVELS

The control factors selected for this project are listed in Table 11.1. Also listed in the table are the control factors' starting and alternate levels. The rationale behind the selection of some of these factors and their levels is given next.

TABLE 11.1 CONTROL FACTORS AND THEIR LEVELS

	Levels*			
Factors	**1**	**2**	**3**	**4**
A. Suction (in of Hg)	1	2		
B. X-Y feed (in/min)	60	80		
C. In-feed (in/min)	10	50		
D. Type of bit	1	2	3	4
E. Spindle position†	1	2	3	4
F. Suction foot	SR	BB		
G. Stacking height (in)	3/16	1/4		
H. Depth of slot (thou)	60	100		
I. Speed (rpm)	30K	40K		

* Starting levels are indicated by an underscore.
† Spindle position is not a control factor. It is a
  noise factor.

Suction (factor A) is used around the router bit to remove the dust as it is generated. Obviously, higher suction could reduce the amount of dust retained on the boards. The starting suction was two inches of mercury (Hg). However, the pump used in the experiment could not produce more suction. So, one inch of Hg was chosen as the alternate level, with the plan that if the experiments showed a significant difference in the dust, a more powerful pump would be obtained.

Related to the suction are *suction foot* and the *depth of the backup slot*. The suction foot determines how the suction is localized near the cutting point. Two types of suction foot (factor F) were chosen: solid ring (SR) and bristle brush (BB). A backup panel is located underneath the panels being routed. Slots are precut in this backup panel to provide air passage and a place for dust to accumulate temporarily. The depth of these slots was a control factor (factor H) in the case study.

Stack height (factor G) and *X-Y feed* (factor B) are control factors related to the productivity of the process—that is, they determine how many boards are cut per hour. The 3/16-in. stack height meant three panels were stacked together while 1/4-in. stack height meant four panels were stacked together. The *in-feed* (factor C) determines the impact force during the lowering of the spindle for starting to cut a new board. Thus, it could influence the life of the bit regarding breakage or damage to its point. Four different *types of router bits* (factor D) made by different manufacturers were investigated in this study. The router bits varied in cutting geometry in terms of the helix angle, the number of flutes, and the type of point.

Spindle position (factor E) is not a control factor. The variation in the state of adjustment of the four spindles is indeed a noise factor for the routing process. All spindle positions must be used in actual production; otherwise, the productivity would suffer. The reason it was included in the study is that in such situations one must choose the settings of control factors that work well with all four spindles. The rationale for including the spindle position along with the control factors is given in Section 11.5.

11.5 DESIGN OF THE MATRIX EXPERIMENT

For this case study, the goal was to not only estimate the main effects of the nine factors listed in the previous section, but also to estimate the four key 2-factor interactions. Note that there are 36 distinct ways of choosing two factors from among nine factors. Thus, the number of two-factor interactions associated with nine factors is 36. An attempt to estimate them all would take excessive experimentation, which is also unnecessary anyway. The four interactions chosen for the case study were the ones judged to be more important based on the knowledge of the cutting process:

1. (X-Y feed) × (speed), that is, B × I
2. (in-feed) × (speed), that is, C × I
3. (stack height) × (speed), that is, G × I
4. (X-Y feed) × (stack height), that is, B × G

The primary purpose of studying interactions in a matrix experiment is to see if any of those interactions are strongly antisynergistic. Lack of strong antisynergistic interactions is the ideal outcome. However, if a strong antisynergistic interaction is found, we should look for a better quality characteristic. If we cannot find such a characteristic, we should look for finding levels of some other control factors that cause the antisynergistic interaction to disappear. Such an approach leads to a stable design compared to an approach where one picks the best combination when a strong antisynergistic interaction is found.

In addition to the requirements listed thus far, the experimenters had to consider the following aspects from a practical viewpoint:

- Suction (factor A) was difficult to change due to difficult access to the pump.

- All four spindles on a machine move in identical ways—that is, they have the same X-Y feed, in-feed, and speed. So, the columns assigned to these factors should be such that groups of four rows can be made, where each group has a common X-Y feed, in-feed, and speed. This allows all four spindles to be used effectively in the matrix experiment.

Construction of the Orthogonal Array

These requirements for constructing the control orthogonal array are fairly complicated. Let us now see how we can apply the advanced strategy described in Chapter 7 to construct an appropriate orthogonal array for this project.

First, the degrees of freedom for this project can be calculated as follows:

Source	Degrees of Freedom
Two 4-level factors	$2 \times (4 - 1) = 6$
Seven 2-level factors	$7 \times (2 - 1) = 7$
Four 2-factor interactions between 2-level columns	$4 \times (2 - 1) \times (2 - 1) = 4$
Overall mean	1
Total	18

Since there are 2-level and 4-level factors in this project, it is preferable to use an array from the 2-level series. Because there are 18 degrees of freedom, the array must have 18 or more rows. The next smallest 2-level array is L_{32}.

The linear graph needed for this case study, called the *required linear graph*, is shown in Figure 11.1(a). Note that each 2-level factor is represented by a dot, and interaction is represented by a line connecting the corresponding dots. Each 4-level factor is represented by two dots, connected by a line according to the column merging method in Chapter 7.

The next step in the advanced strategy for constructing orthogonal arrays is to select a suitable linear graph of the orthogonal array L_{32} and modify it to fit the required linear graph. Here we take a slightly different approach. We first simplify the required linear graph by taking advantage of the special circumstances and then proceed to fit a standard linear graph.

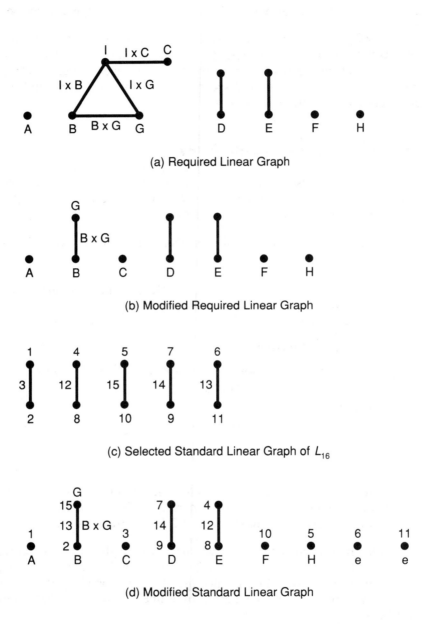

Figure 11.1 Linear graph for the routing project.

We notice that we must estimate the interactions of the factor I with three other factors. One way to simplify the required linear graph is to treat I as an *outer factor*—that is, to first construct an orthogonal array by ignoring I and its interactions.

Then, conduct each row of the orthogonal array with the two levels of I. By so doing, we can estimate the main effect of I and also the interactions of I with all other factors.

The modified required linear graph, after dropping factor I and its interactions is shown in Figure 11.1(b). This is a much simpler linear graph and, hence, easier to fit to a standard linear graph. Dropping the 2-level factor I and its interactions with three, 2-level factors is equivalent to reducing the degrees of freedom by four. Thus, there are 14 degrees of freedom associated with the linear graph of Figure 11.1(b). Therefore, the orthogonal array L_{16} can be used to fit the linear graph of Figure 11.1(b). This represents a substantial simplification compared with having to use the array L_{32} for the original required linear graph. The linear graph of Figure 11.1(b) has three lines, connecting two dots each, and four isolated dots. Thus, a standard linear graph that has a number of lines that connect pairs of dots seems most appropriate. Such a linear graph was selected from the standard linear graphs of L_{16} given in Appendix C and it is shown in Figure 11.1(c). It has five lines, each connecting two distinct dots. The step-by-step modification of this linear graph to make it fit the one in Figure 11.1(b) is discussed next.

The requirement that there should be as few changes as possible in factor A implies that factor A should be assigned to column 1. Therefore, we break the line connecting dots 1 and 2, giving us three dots 1, 2, and 3. Now, we want to make groups of four rows which have common X-Y feed (factor B) and in-feed (factor C). Therefore, these two factors should be assigned to columns that have fewer changes. Hence, we assign columns 2 and 3 to factors B and C, respectively. Referring to the columns 2 and 3 of the array L_{16} in Table 11.2, it is clear that each of the four groups of four rows (1-4, 5-8, 9-12, and 13-16) has a common X-Y feed and in-feed. Now we should construct a 4-level column for spindle position (factor E) so that all four spindle positions will be present in each group of the four rows mentioned above. We observe that this can be achieved by merging columns 4 and 8 according to Section 7.7. Of course, column 12, which represents the interaction between columns 4 and 8, must be kept empty. Note that we could have used any of the other three lines in Figure 11.1(c) for this purpose.

Next, from among the remaining three lines in the standard linear graph, we arbitrarily chose columns 7 and 9 to form a 4-level column for factor D. Of course, the interaction column 14 must be kept empty.

The two remaining lines are then broken to form six isolated dots corresponding to columns 5, 10, 15, 6, 11, and 13. The next priority is to pick a column for factor G so that the interaction B × G would be contained in one of the remaining five columns. For this purpose, we refer to the interaction table for the L_{16} array given in Appendix C. We picked column 15 for factor G. Column 13 contains interaction between columns 2 and 15, so it can be used to estimate the interaction B × G. We indicate this in the linear graph by a line joining the dots for the columns 2 and 15.

From the remaining four columns, we arbitrarily assign columns 10 and 5 to factors F and H. Columns 6 and 11 are kept empty.

TABLE 11.2 L_{16} ORTHOGONAL ARRAY

Expt. No.	Column Number														
	1	2	3	4	5	6	7	8	9	10	11	12	13	14	15
1	1	1	1	1	1	1	1	1	1	1	1	1	1	1	1
2	1	1	1	1	1	1	1	2	2	2	2	2	2	2	2
3	1	1	1	2	2	2	2	1	1	1	1	2	2	2	2
4	1	1	1	2	2	2	2	2	2	2	2	1	1	1	1
5	1	2	2	1	1	2	2	1	1	2	2	1	1	2	2
6	1	2	2	1	1	2	2	2	2	1	1	2	2	1	1
7	1	2	2	2	2	1	1	1	1	2	2	2	2	1	1
8	1	2	2	2	2	1	1	2	2	1	1	1	1	2	2
9	2	1	2	1	2	1	2	1	2	1	2	1	2	1	2
10	2	1	2	1	2	1	2	2	1	2	1	2	1	2	1
11	2	1	2	2	1	2	1	1	2	1	2	2	1	2	1
12	2	1	2	2	1	2	1	2	1	2	1	1	2	1	2
13	2	2	1	1	2	2	1	1	2	2	1	1	2	2	1
14	2	2	1	1	2	2	1	2	1	1	2	2	1	1	2
15	2	2	1	2	1	1	2	1	2	2	1	2	1	1	2
16	2	2	1	2	1	1	2	2	1	1	2	1	2	2	1

The final, modified standard linear graph along with the assignment of factors to columns is shown in Figure 11.1(d). The assignment of factors to the columns of the L_{16} array is as follows:

Factor	Column	Factor	Column
A	1	F	10
B	2	G	15
C	3	H	5
D	7, 9, 14	B × G	13
E	4, 8, 12		

The 4-level columns for factors D and E were formed in the array L_{16} by the column merging method of Section 7.7. The resulting 16 row orthogonal array is the same as the first 16 rows of Table 11.3, except for the column for factor I. Because I is an outer factor, we obtain the entire matrix experiment as follows: make rows 17-32 the same as rows 1-16. Add a column for factor I that has 1 in the rows 1-16 and 2 in the rows 17-32. Note that the final matrix experiment shown in Table 11.3 is indeed an orthogonal array—that is, in every pair of columns, all combinations occur and they occur an equal number of times. We ask the reader to verify this claim for a few pairs of columns. Note that the matrix experiment of Table 11.3 satisfies all the requirements set forth earlier in this section.

The 32 experiments in the control orthogonal array of Table 11.3 are arranged in eight groups of four experiments such that:

a. For each group there is a common speed, X-Y feed, and in-feed

b. The four experiments in each group correspond to four different spindles

Thus, each group constitutes a machine run using all four spindles, and the 32 experiments in the control orthogonal array can be conducted in eight runs of the routing machine.

Observe the ease with which we were able to construct an orthogonal array for a very complicated combinatoric problem using the standard orthogonal arrays and linear graphs prepared by Taguchi.

Inclusion of a Noise Factor in a Matrix Experiment

As a rule, noise factors should not be mixed in with the control factors in a matrix experiment (orthogonal array experiment). Instead, noise factors should be used to form different testing conditions so that the S/N ratio can accurately measure sensitivity to noise factors. According to this rule, we should have dropped the spindled position column in Table 11.3 and considered the four spindle positions as four testing conditions for each row of the orthogonal array, which would amount to a four times larger experimental effort.

To save the experimental effort, in some situations we assign control factors as well as noise factors to the columns of the matrix experiment. In the router bit life improvement case study, we included the spindle position in the matrix experiment for the same reason. Note that in such matrix experiments noise is introduced systematically and in a balanced manner. Also, results from such matrix experiments are more dependable than the results from a matrix experiment where noise is fixed at one level to save experimental effort. Control factor levels found optimum through such experiments are preferred levels, on average, for all noise factor levels in the experiment. In the router bit example, results imply that the control factor levels found optimum are, on average, preferred levels for all four spindle positions.

TABLE 11.3 MATRIX EXPERIMENT AND OBSERVED LIFE

Expt. No.	Suction A	X-Y Feed B	In-feed C	Bit D	Spindle Position E	Suction Foot F	Stack Height G	Depth H	Speed I	Observed Life*
1	1	1	1	1	1	1	1	1	1	3.5
2	1	1	1	2	2	2	2	1	1	0.5
3	1	1	1	3	4	1	2	2	1	0.5
4	1	1	1	4	3	2	1	2	1	17.5
5	1	2	2	3	1	2	2	1	1	0.5
6	1	2	2	4	2	1	1	1	1	2.5
7	1	2	2	1	4	2	1	2	1	0.5
8	1	2	2	2	3	1	2	2	1	0.5
9	2	1	2	4	1	1	2	2	1	17.5
10	2	1	2	3	2	2	1	2	1	2.5
11	2	1	2	2	4	1	1	1	1	0.5
12	2	1	2	1	3	2	2	1	1	3.5
13	2	2	1	2	1	2	1	2	1	0.5
14	2	2	1	1	2	1	2	2	1	2.5
15	2	2	1	4	4	2	2	1	1	0.5
16	2	2	1	3	3	1	1	1	1	3.5
17	1	1	1	1	1	1	1	1	2	17.5
18	1	1	1	2	2	2	2	1	2	0.5
19	1	1	1	3	4	1	2	2	2	0.5
20	1	1	1	4	3	2	1	2	2	17.5
21	1	2	2	3	1	2	2	1	2	0.5
22	1	2	2	4	2	1	1	1	2	17.5
23	1	2	2	1	4	2	1	2	2	14.5
24	1	2	2	2	3	1	2	2	2	0.5
25	2	1	2	4	1	1	2	2	2	17.5
26	2	1	2	3	2	2	1	2	2	3.5
27	2	1	2	2	4	1	1	1	2	17.5
28	2	1	2	1	3	2	2	1	2	3.5
29	2	2	1	2	1	2	1	2	2	0.5
30	2	2	1	1	2	1	2	2	2	3.5
31	2	2	1	4	4	2	2	1	2	0.5
32	2	2	1	3	3	1	1	1	2	17.5

* Life was measured in hundreds of inches of movement in X-Y plane. Tests were terminated at 1,700 inches.

11.6 EXPERIMENTAL PROCEDURE

In order to economize on the size of the experiment, the experimenters took only one observation of router bit life per row of the control orthogonal array. Of course, they realized that taking two or three noise conditions per row of the control orthogonal array would give them more accurate conclusions. However, doing this would mean exceeding the allowed time and budget. Thus, a total of only 32 bits were used in this project to determine the optimum settings of the control factors.

During each machine run, the machine was stopped after every 100 in. of cut (that is, 100 in. of router bit movement in the X-Y plane) to inspect the amount of dust. If the dust was beyond a certain minimum predetermined level, the bit was recorded as failed. Also, if a bit broke, it was obviously considered to have failed. Otherwise, it was considered to have survived.

Before the experiment was started, the average bit life was around 850 in. Thus, each experiment was stopped at 1,700 in. of cut, which is twice the original average life, and the survival or failure of the bit was recorded.

Usually, measuring the exact failure time of a product is very difficult. Therefore, in practice, it is preferable to make periodic checks for survival as was done every 100 in. of cut in this case study. Also, running a life test beyond a certain point costs a lot, and it does not add substantially to the information about the preferred level of a control factor. Therefore, truncating the life test at an appropriate point is recommended. In the reliability analysis or reliability engineering literature, determining an interval in which a product fails is called *interval censoring*, whereas terminating a life test at a certain point is called *right censoring*.

11.7 DATA ANALYSIS

Table 11.3 gives the experimental data in hundreds of inches. A reading of 0.5 means that the bit failed prior to the first inspection at 100 in. A reading of 3.5 means that the bit failed between 300 and 400 in. Other readings have similar interpretation, except the reading of 17.5 which means survival beyond 1,700 in., the point where the experiment was terminated. Notice that for 14 experiments, the life is 0.5 (50 in.), meaning that those conditions are extremely unfavorable. Also, there are eight cases of life equal to 17.5, which are very favorable conditions. *During experimentation, it is important to take a broad range for each control factor so that a substantial number of favorable and unfavorable conditions are created.* Much can be learned about the optimum settings of control factors when there is such diversity of data.

Now we will show two simple and separate analyses of the life data for determining the best levels for the control factors. The first analysis is aimed at determining the effect of each control factor on the mean failure time. The second analysis, described in the next section, is useful for determining the effect of changing the level of each factor on the survival probability curve.

The life data was analyzed by the standard procedures described in Chapter 3 to determine the effects of the control factors on the mean life. The mean life for each factor level and the results of analysis of variance are given in Table 11.4. These results are plotted in Figure 11.2. Note that in this analysis we have ignored the effect of both types of censoring. The following conclusions are apparent from the plots in Figure 11.2:

- 1-in. suction is as good as 2-in. suction. Therefore, it is unnecessary to increase suction beyond 2 in.

- Slower X-Y feed gives longer life.

- The effect of in-feed is small.

TABLE 11.4 FACTOR EFFECTS AND ANALYSIS OF VARIANCE FOR ROUTER BIT LIFE*

Factor	Level Means†				Sum of Squares	Degrees of Freedom	Mean Square	F
	1	2	3	4				
A. Suction	5.94	5.94			0	1	0	0.0
B. X-Y feed	7.75	4.13			105.13	1	105.13	4.1
C. In-feed	5.44	6.44			8.00	1	8.00	0.3
D. Type of bit	6.03	2.63	3.63	11.38	367.38	3	122.46	4.8
E. Spindle position	7.25	4.13	8.00	4.38	93.63	3	31.21	1.2
F. Suction foot	7.69	4.19			98.00	1	98.00	3.9
G. Stack height	8.56	3.31			220.50	1	220.50	8.7
H. Depth of slot	5.63	6.25			3.13	1	3.13	0.1
I. Speed	3.56	8.31			180.50	1	180.50	7.1
I × B					10.50	1	10.50	0.4
I × C					10.13	1	10.13	0.4
I × G					171.13	1	171.13	6.7
B × G					4.50	1	4.50	0.2
Error					355.37	14	25.38	
Total					1627.90	31		

* Life in hundreds of inches.
† Overall mean life = 5.94 hundreds of inches; starting conditions are identified by an underscore.

- The starting bit is the best of the four types.

- The difference among the spindles is small. However, we should check the centering of spindles 2 and 4.

- Solid ring suction foot is better than the existing bristle brush type.

- Lowering the stack height makes a large improvement. This change, however, raises a machine productivity issue.

- The depth of the slot in the backup material has a negligible effect.

- Higher rotational speed gives improved life. If the machine stability permits, even higher speed should be tried in the next cycle of experiments.

- The only 2-factor interaction that is large is the stack height versus speed interaction. However, this interaction is of the synergistic type. Therefore, the optimum settings of these factors suggested by the main effects are consistent with those suggested by the interaction.

Optimum Control Factor Settings

The best settings of the control factors, called *optimum 1*, suggested by the results above, along with their starting levels, are displayed side by side in Table 11.5.

Using the linear model and taking into consideration only the terms for which the variance ratio is large (that is, the factors B, D, F, G, I and interaction I × G), we can predict the router bit life under the starting, optimum, or any other combination of control factor settings. The predicted life under the starting and optimum conditions are 888 in. and 2,225 in., respectively. The computations involved in the prediction are displayed in Table 11.5. Note that the contribution of the I × G interaction under starting conditions was computed as follows:

(Contribution at I_2, G_2 due to I × G interaction)

$$= (m_{I_2 G_2} - \mu) - (m_{I_2} - \mu) - (m_{G_2} - \mu)$$

$$= (3.38 - 5.94) - (8.31 - 5.94) - (3.31 - 5.94)$$

$$= -2.30$$

Note that $m_{I_2 G_2}$ is the mean life for the experiments with speed I_2 and stack height G_2. The contribution of the I × G interaction under the optimum conditions was computed in a similar manner. Because of the censoring during the experiment at 1,700

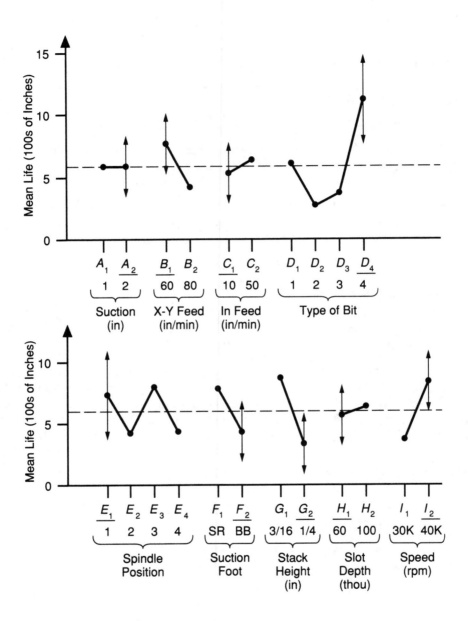

Figure 11.2 Main effects of control factors on router bit life and some 2-factor interactions. Two-standard-deviation confidence limits on the main effect for the starting level are also shown.

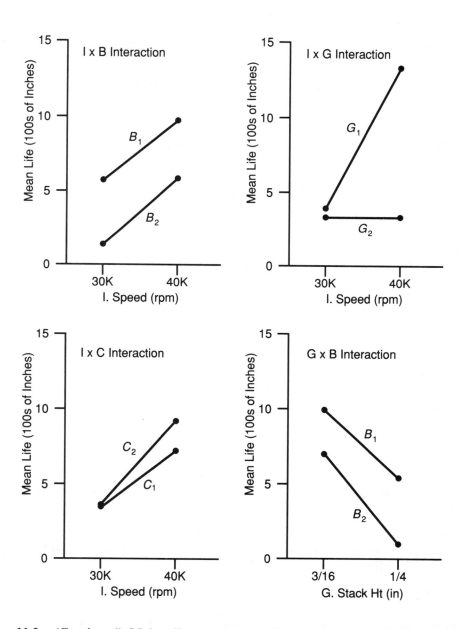

Figure 11.2   (Continued) Main effects of control factors on router bit life and some 2-factor interactions.

TABLE 11.5 PREDICTION OF LIFE USING ADDITIVE MODEL*

Factor	Starting Condition		Optimum 1		Optimum 2	
	Setting	Contribution	Setting	Contribution	Setting	Contribution
A. Suction	A_2	–	A_2	–	A_2	–
B. X-Y feed	B_1	1.81	B_1	1.81	B_2	–1.81
C. In-feed	C_1	–	C_1	–	C_1	–
D. Type of bit	D_4	5.44	D_4	5.44	D_4	5.44
E. Spindle position	E_1–E_4	–	E_1–E_4	–	E_1–E_4	–
F. Suction foot	F_2	–1.75	F_1	1.75	F_1	1.75
G. Stack height	G_2	–2.63	G_1	2.63	G_1	2.63
H. Depth of slot	H_1	–	H_1	–	H_1	–
I. Speed	I_2	2.37	I_2	2.37	I_2	2.37
I × G interaction	I_2G_2	–2.30	I_2G_1	2.31	I_2G_1	2.31
Overall mean		5.94		5.94		5.94
Total		8.88		22.25		18.63

* Life in hundreds of inches.

in., these predictions, obviously, are likely to be on the low side, especially the prediction under optimum conditions which is likely to be much less than the realized value.

From the machine logs, the router bit life under starting conditions was found to be 900 in., while the verification (confirmatory) experiment under optimum conditions yielded an average life in excess of 4,150 in.

In selecting the best operating conditions for the routing process, one must consider the overall cost, which includes not only the cost of router bits but also the cost of machine productivity, the cost of cleaning the boards if needed, etc. Under the optimum conditions shown in Table 11.5, the stack height is 3/16 in. as opposed to 1/4 in. under the starting conditions. This means three panels are cut simultaneously instead of four panels. However, the lost machine productivity caused by this change can be made up by increasing the X-Y feed. If the X-Y feed is increased to 80 in. per minute, the productivity of the machine would get back approximately to the starting level. The predicted router bit life under these alternate optimum conditions, called *optimum 2*, is 1,863 in., which is about twice the predicted life for starting conditions. Thus, a 50-percent reduction in router bit cost can be achieved while still maintaining machine productivity. An auxiliary experiment typically would be needed to estimate precisely the effect of X-Y feed under the new settings of all other factors. This would enable us to make an accurate economic analysis.

In summary, orthogonal array based matrix experiments are useful for finding optimum control factor settings with regard to product life. In the router bit example, the experimenters were able to improve the router bit life by a factor of 2 to 4.

Accelerated Life Tests

Sometimes, in order to see any failures in a reasonable time, life tests must be conducted under stressed conditions, such as higher than normal temperature or humidity. Such life tests are called *accelerated life tests*. An important concern in using accelerated life tests is how to ensure that the control factor levels found optimum during the accelerated tests will also be optimum under normal conditions. This can be achieved by including several stress levels in the matrix experiment and demonstrating additivity. For an application of the Robust Design method for accelerated life tests, see Phadke, Swann, and Hill [P6] and Mitchell [M1].

11.8 SURVIVAL PROBABILITY CURVES

The life data can also be analyzed in a different way (refer to the *minute analysis method* described in Taguchi [T1] and Taguchi and Wu [T7]) to construct the survival probability curves for the levels of each factor. To do so, we look at every 100 in. of cut and note which router bits failed and which survived. Table 11.6 shows the survival data displayed in this manner. Note that a 1 means survival and a 0 means failure. The survival data at every time point can be analyzed by the standard method described in Chapter 3 to determine the effects of various factors. Thus, for suction levels A_1 and A_2, the level means at 100 in. of cut are 0.4375 and 0.6875, respectively. These are nothing but the fraction of router bits surviving at 100 in. of cut for the two levels of suction. The survival probabilities can be estimated in a similar manner for each factor and each time period—100 in., 200 in., etc. These data are plotted in Figure 11.3. These plots graphically display the effects of factor level changes on the entire life curve and can be used to decide the optimum settings of the control factors. In this case, the conclusions from these plots are consistent with those from the analysis described in Section 11.6.

Plots similar to those in Figure 11.3 can be used to predict the entire survival probability curve under a new set of factor level combinations such as the optimum combination. The prediction method is described in Chapter 5, Section 5.5 in conjunction with the analysis of ordered categorical data (see also Taguchi and Wu [T7].

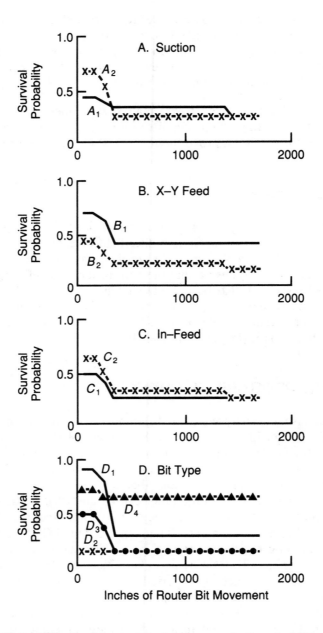

Figure 11.3   Effects of control factors on survival probability curves.

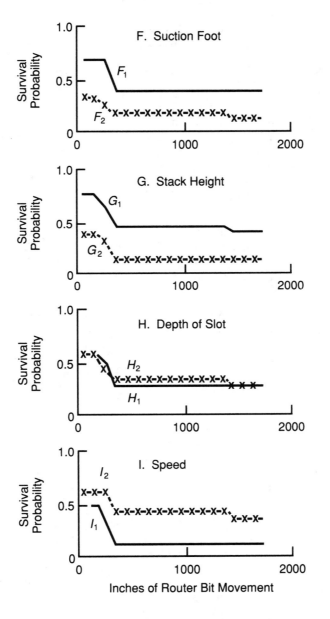

Figure 11.3   (Continued) Effects of control factors on survival probability curves.

TABLE 11.6 SURVIVAL DATA AT VARIOUS TIME POINTS

Expt. No.	Survival at Various Time Points* (100s of inches)																
	1.	2.	3.	4.	5.	6.	7.	8.	9.	10.	11.	12.	13.	14.	15.	16.	17.
1	1	1	1	0	0	0	0	0	0	0	0	0	0	0	0	0	0
2	0	0	0	0	0	0	0	0	0	0	0	0	0	0	0	0	0
3	0	0	0	0	0	0	0	0	0	0	0	0	0	0	0	0	0
4	1	1	1	1	1	1	1	1	1	1	1	1	1	1	1	1	1
5	0	0	0	0	0	0	0	0	0	0	0	0	0	0	0	0	0
6	1	1	0	0	0	0	0	0	0	0	0	0	0	0	0	0	0
7	0	0	0	0	0	0	0	0	0	0	0	0	0	0	0	0	0
8	0	0	0	0	0	0	0	0	0	0	0	0	0	0	0	0	0
9	1	1	1	1	1	1	1	1	1	1	1	1	1	1	1	1	1
10	1	1	0	0	0	0	0	0	0	0	0	0	0	0	0	0	0
11	0	0	0	0	0	0	0	0	0	0	0	0	0	0	0	0	0
12	1	1	1	0	0	0	0	0	0	0	0	0	0	0	0	0	0
13	0	0	0	0	0	0	0	0	0	0	0	0	0	0	0	0	0
14	1	1	0	0	0	0	0	0	0	0	0	0	0	0	0	0	0
15	0	0	0	0	0	0	0	0	0	0	0	0	0	0	0	0	0
16	1	1	1	0	0	0	0	0	0	0	0	0	0	0	0	0	0
17	1	1	1	1	1	1	1	1	1	1	1	1	1	1	1	1	1
18	0	0	0	0	0	0	0	0	0	0	0	0	0	0	0	0	0
19	0	0	0	0	0	0	0	0	0	0	0	0	0	0	0	0	0
20	1	1	1	1	1	1	1	1	1	1	1	1	1	1	1	1	1
21	0	0	0	0	0	0	0	0	0	0	0	0	0	0	0	0	0
22	1	1	1	1	1	1	1	1	1	1	1	1	1	1	1	1	1
23	1	1	1	1	1	1	1	1	1	1	1	1	1	1	0	0	0
24	0	0	0	0	0	0	0	0	0	0	0	0	0	0	0	0	0
25	1	1	1	1	1	1	1	1	1	1	1	1	1	1	1	1	1
26	1	1	1	0	0	0	0	0	0	0	0	0	0	0	0	0	0
27	1	1	1	1	1	1	1	1	1	1	1	1	1	1	1	1	1
28	1	1	1	0	0	0	0	0	0	0	0	0	0	0	0	0	0
29	0	0	0	0	0	0	0	0	0	0	0	0	0	0	0	0	0
30	1	1	1	0	0	0	0	0	0	0	0	0	0	0	0	0	0
31	0	0	0	0	0	0	0	0	0	0	0	0	0	0	0	0	0
32	1	1	1	1	1	1	1	1	1	1	1	1	1	1	1	1	1

* Entry 1 means survival of the bit; entry 0 means failure.

Note that in this method of determining life curves, no assumption was made regarding the shape of the curve—such as Weibull or log-normal distribution. Also, the total amount of data needed to come up with the life curves is small. In this example, it took only 32 samples to determine the effects of eight control factors. For a single good fit of a Weibull distribution, one typically needs several tens of observations. So, the approach used here can be very beneficial for reliability improvement projects.

The approach of determining survival probability curves described here is similar to the Eulerian method of analyzing fluid flows. In the Eulerian method, one looks at fixed boundaries in space and examines the fluid masses crossing those boundaries. In the life study example, we look at fixed time points (here, measured in inches of cut) and observe which samples survive past those time points. The survival probability curves are constructed from these observations. In the Lagrangian method of analyzing fluid flows, one tracks a fluid particle and examines the changes in velocities, pressure, etc., experienced by the particle. Fluid flow equations are derived from this examination. In the reliability study, the analogous way would be to observe the actual failure times of each sample and then analyze the failure data. The analysis of Section 11.6 was based on this approach. However, when there are many censored observations (in this example eight router bits did not fail by the time the experiment was stopped and readings were taken only at 100-in. intervals), the approach of this section, where survival probability curves are estimated, gives more comprehensive information.

11.9 SUMMARY

- Reliability improvement can be accomplished during the design stage in one of the three ways: (1) reduce sensitivity of the product's function to the variation in product parameters, (2) reduce the rate of change of the product parameters, and (3) include redundancy. The first way is most cost-effective and it is the same as parameter design. The second way is analogous to tolerance design and it involves using more expensive components or manufacturing processes. The third way is used when the cost of failure is very large compared to the cost of providing spare components or the whole product.

- Finding an appropriate continuous quality characteristic and reducing its sensitivity to noise factors is the best way of improving reliability. When an appropriate quality characteristic cannot be found, then only life tests should be considered for life improvement. Matrix experiments using orthogonal arrays can be used to conduct life tests efficiently with several control factors.

- The S/N ratio calculated from a continuous quality characteristic can be used to estimate the average life of a new product. Let η_1 be the S/N ratio for the new product and η_2 be the S/N ratio for a benchmark product whose average life is known. Then the average life of the new product is r times the average life of the benchmark product, where

$$r = 10^{\left(\frac{\eta_1 - \eta_2}{10}\right)}.$$

- The goal of the router bit case study was to reduce dust formation. Since there existed no continuous quality characteristic that could be observed conveniently, the life test was conducted to improve the router bit life.

- Effects of nine factors, eight control factors and spindle position, were studied using an orthogonal array with 32 experiments. Out of the nine factors, two factors had four levels, and the remaining seven had two levels. Four specific 2-factor interactions were also studied. In addition, there were several physical restrictions regarding the factor levels. Use of Taguchi's linear graphs made it easy to construct the orthogonal array, which allowed the estimation of desired factor main effects and interactions while satisfying the physical restrictions.

- Only one router bit was used per experiment. Dust formation was observed every 100 in. of cut in order to judge the failure of the bit. The length of cut prior to formation of appreciable dust or breakage of the bit was called the bit life and it was used as the quality characteristic. Each experiment was terminated at 1,700 in. of cut regardless of failure or survival of the bit. Thus the life data were censored.

- Effects of the nine factors on router bit life were computed and optimum levels for the control factors were identified. Under a set of optimum conditions, called *optimum 1*, a 4-fold increase in router bit life was observed, but with a 12.5-percent reduction in throughput. Under another set of optimum conditions, called *optimum 2*, a 2-fold increase in router bit life was observed, with no drop in throughput.

- The life data from a matrix can also be analyzed by the minute analysis method to determine the effects of the control factors on the survival probability curves. This method of analysis does not presume any failure time distribution, such as log-normal or Weibull distribution. Also, the total amount of data needed to determine the survival probability curves is small.

Appendix A

ORTHOGONALITY OF A MATRIX EXPERIMENT

Before defining the orthogonality of a matrix experiment we need to review the following definitions from linear algebra and statistics. Recall that η_1, η_2, ..., η_9 are the observations for the nine rows of the matrix experiment given by Table 3.2. Consider the *linear form, L_i,* given by

$$L_i = w_{i1}\eta_1 + w_{i2}\eta_1 + \cdots + w_{i9}\eta_9 \qquad (A.1)$$

which is a weighted sum of the nine observations. The linear form L_i is called a *contrast* if the weights add up to zero—that is, if

$$w_{i1} + w_{i2} + \cdots + w_{i9} = 0 \qquad (A.2)$$

Two contrasts, L_1 and L_2, are said to be orthogonal if the inner product of the vectors corresponding to their weights is zero—that is, if

$$w_{11}w_{21} + w_{12}w_{22} + \cdots + w_{19}w_{29} = 0 \qquad (A.3)$$

Let us consider three weights w_{11}, w_{12}, and w_{13} corresponding to the three levels in the column 1 of the matrix experiment given by Table 3.2 (Chapter 3). Then we call the following linear form, L_1, the contrast corresponding to the column 1

$$L_1 = w_{11}\eta_1 + w_{11}\eta_2 + w_{11}\eta_3 + w_{12}\eta_4 + w_{12}\eta_5 + w_{12}\eta_6$$

$$+ w_{13}\eta_7 + w_{13}\eta_8 + w_{13}\eta_9 \qquad (A.4)$$

provided all weights add up to zero. In this case, it implies

$$w_{11} + w_{12} + w_{13} = 0. \qquad (A.5)$$

Note that in Equation (A.4) we use the weight w_{11} whenever the level is 1, weight w_{12} whenever the level is 2, and weight w_{13} whenever the level is 3.

An array used in a matrix experiment is called an orthogonal array if the contrasts corresponding to all its columns are mutually orthogonal. Let us consider columns 1 and 2 of the matrix experiment given by Table 3.2. Equation (A.4) is the contrast corresponding to column 1. Let w_{21}, w_{22}, and w_{23} be the weights corresponding to the three levels in column 2. Then, the contrast corresponding to column 2 is L_2, where

$$L_2 = w_{21}\eta_1 + w_{22}\eta_2 + w_{23}\eta_3 + w_{21}\eta_4 + w_{22}\eta_5 + w_{23}\eta_6$$

$$+ w_{21}\eta_7 + w_{22}\eta_8 + w_{23}\eta_9. \qquad (A.6)$$

Of course, the weights in Equation (A.6) must add up to zero—that is

$$w_{21} + w_{22} + w_{23} = 0. \qquad (A.7)$$

The inner product of the vectors corresponding to the weights in the two contrasts L_1 and L_2 is given by

$$w_{11}w_{21} + w_{11}w_{22} + w_{11}w_{23} + w_{12}w_{21} + w_{12}w_{22} + w_{12}w_{23}$$

$$+ w_{13}w_{21} + w_{13}w_{22} + w_{13}w_{23}$$

$$= \left[w_{11} + w_{12} + w_{13} \right] \left[w_{21} + w_{22} + w_{23} \right]$$

$$= 0.$$

Hence, columns 1 and 2 are mutually orthogonal. The orthogonality of all pairs of columns in the matrix experiment given by Table 3.2 can be verified in a similar manner. In general it can be shown that the balancing property is a sufficient condition for a matrix experiment to be orthogonal.

In column 2 of the matrix experiment given by Table 3.2, suppose we replace level 3 by level 2. Then looking at columns 1 and 2 it is clear that the two columns do not have the balancing property. However, for level 1 in column 1 there is one row with level 1 in column 2 and two rows with level 2 in column 2. Similarly, for level 2 in column 1, there is one row with level 1 in column 2 and two rows with level 2 in column 2. The same can be said about level 3 in column 1. This is called *proportional balancing*. It can be shown that proportional balancing is also a sufficient condition for a matrix experiment to be orthogonal.

Among the three weights corresponding to column 1, we can independently choose the values of any two of them. The third weight is then determined by Equation (A.5). Hence, we say that column 1 has two degrees of freedom. In general, a column with n levels has $n - 1$ degrees of freedom.

Appendix B

UNCONSTRAINED OPTIMIZATION

Here we define in precise mathematical terms the problem of minimizing the variance of thickness while keeping the mean on target and derive its solution.

Let $\mathbf{z} = (z_1, z_2, \ldots, z_q)^T$ be the vector formed by the control factors; $\mathbf{x}$ be the vector formed by the noise factors; and $y(\mathbf{x}; \mathbf{z})$ denote the observed quality characteristic, namely the polysilicon layer thickness, for particular values of the noise and control factors. Note that y is nonnegative. Let $\mu(\mathbf{z})$ and $\sigma^2(\mathbf{z})$ denote the mean and variance of the response. Obviously, μ and σ^2 are obtained by integrating with respect to the probability density of $\mathbf{x}$, and hence are functions of only $\mathbf{z}$.

Problem Statement:

The optimization problem can be stated as follows:

$$\text{Minimize} \quad \sigma^2(\mathbf{z})$$
$$\mathbf{z} \tag{B.1}$$
$$\text{Subject to} \quad \mu(\mathbf{z}) = \mu_0 \ .$$

This is a constrained optimization problem and very difficult to solve experimentally.

Solution:

We postulate that one of the control factors is a scaling factor, say, z_1. It, then, implies that

$$y(\mathbf{x}; \mathbf{z}) = z_1 h(\mathbf{x}; \mathbf{z}') \tag{B.2}$$

for all $\mathbf{x}$ and $\mathbf{z}$, where $\mathbf{z}' = (z_2, z_3, \ldots, z_q)^T$, and $h(\mathbf{x}; \mathbf{z}')$ does not depend on z_1.

It follows that

$$\mu(\mathbf{z}) = z_1 \, \mu_h(\mathbf{z}') \tag{B.3}$$

and

$$\sigma^2(\mathbf{z}) = z_1^2 \, \sigma_h^2(\mathbf{z}') \tag{B.4}$$

where μ_h and σ_h^2 are, respectively, the mean and variance of $h(\mathbf{x}, \mathbf{z})$.

Suppose $\mathbf{z}^* = (z_1^*, \mathbf{z}'^*)$ is chosen by the following procedure:

(a) $\mathbf{z}'^*$ is an argument that minimizes $\dfrac{\sigma_h^2(\mathbf{z}')}{\mu_h^2(\mathbf{z}')}$, and

(b) z_1^* is chosen such that $\mu(z_1^*, \mathbf{z}'^*) = \mu_0$.

We will now show that $\mathbf{z}^*$ is an optimum solution to the problem defined by Equation (B.1).

First, note that $\mathbf{z}^*$ is a feasible solution since $\mu(z_1^*, \mathbf{z}'^*) = \mu_0$. Next consider any feasible solution $\mathbf{z} = (z_1, \mathbf{z}')$. We have

$$\sigma^2(z_1, \mathbf{z}') = \mu^2(z_1, \mathbf{z}') \cdot \frac{\sigma^2(z_1, \mathbf{z}')}{\mu^2(z_1, \mathbf{z}')}$$

$$= \mu_0^2 \cdot \frac{\sigma_h^2(\mathbf{z}')}{\mu_h^2(\mathbf{z}')} \cdot \tag{B.5}$$

Combining the definition of z'^* in (a) above and Equation (B.5), we have for all feasible solutions,

$$\sigma^2(z_1, z') \geq \sigma^2(z_1*, z'^*).$$

Thus, $z^* = (z_1, z'^*)$ is an optimal solution for the problem defined by Equation (B.1).

Referring to Equations (B.3) and (B.4) it is clear that

$$\frac{\sigma^2(z_1, z')}{\mu^2(z_1, z')} = \frac{\sigma_h^2(z')}{\mu_h^2(z')}$$

for any value of z_1. Therefore, step (a) can be solved as follows:

(a')   choose any value for z_1, then

(a'')   find z'^* that minimizes $\dfrac{\sigma^2(z_1, z')}{\mu^2(z_1, z')}$

In fact, it is not even necessary to know which control factor is a scaling factor. We can discover the scaling factor by examining the effects of all control factors on the signal-to-noise (S/N) ratio and the mean. Any factor that has no effect on the S/N ratio, but a significant effect on the mean can be used as a scaling factor.

In summary, the original constrained optimization problem can be solved as an unconstrained optimization problem, step (a), followed by adjusting the mean on target, step (b). For obvious reasons, this procedure is called a 2-step procedure. For further discussions on the 2-step procedure, see Taguchi and Phadke [T6]; Phadke and Dehnad [P4]; Leon, Shoemaker and Kackar [L2]; Nair and Pregibon [N2]; and Box [B1]. The particular derivation given in this appendix was suggested by M. Hamami.

Note that the derivation above is perfectly valid if we replace z_1 in Equation (B.2) by an arbitrary function $g(z_1)$ of z_1. This represents a generalization of the common concept of linear scaling.

Appendix C

STANDARD ORTHOGONAL
ARRAYS
AND
LINEAR GRAPHS*

---

* The orthogonal arrays and linear graphs are reproduced with permission from Dr. Genichi Taguchi and with help from Mr. John Kennedy of American Supplier Institute, Inc. For more details of the orthogonal arrays and linear graphs, see Taguchi [T1] and Taguchi and Konishi [T5].

$L_4 (2^3)$ Orthogonal Array

Expt. No.	Column 1	2	3
1	1	1	1
2	1	2	2
3	2	1	2
4	2	2	1

Linear Graph for L_4

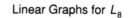

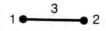

$L_8 (2^7)$ Orthogonal Array

Expt. No.	Column 1	2	3	4	5	6	7
1	1	1	1	1	1	1	1
2	1	1	1	2	2	2	2
3	1	2	2	1	1	2	2
4	1	2	2	2	2	1	1
5	2	1	2	1	2	1	2
6	2	1	2	2	1	2	1
7	2	2	1	1	2	2	1
8	2	2	1	2	1	1	2

Interaction Table for L_8

Column	Column 1	2	3	4	5	6	7
1	(1)	3	2	5	4	7	6
2		(2)	1	6	7	4	5
3			(3)	7	6	5	4
4				(4)	1	2	3
5					(5)	3	2
6						(6)	1
7							(7)

Linear Graphs for L_8

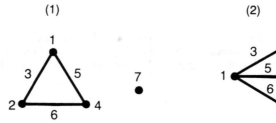

(1)   (2)

$$L_9 \ (3^4)$$

$L_9 \ (3^4)$ Orthogonal Array

Expt. No.	Column 1	2	3	4
1	1	1	1	1
2	1	2	2	2
3	1	3	3	3
4	2	1	2	3
5	2	2	3	1
6	2	3	1	2
7	3	1	3	2
8	3	2	1	3
9	3	3	2	1

Linear Graphs for L_9

$$L_{12} \ (2^{11})$$

$L_{12} \ (2^{11})$ Orthogonal Array

Expt. No.	Column 1	2	3	4	5	6	7	8	9	10	11
1	1	1	1	1	1	1	1	1	1	1	1
2	1	1	1	1	1	2	2	2	2	2	2
3	1	1	2	2	2	1	1	1	2	2	2
4	1	2	1	2	2	1	2	2	1	1	2
5	1	2	2	1	2	2	1	2	1	2	1
6	1	2	2	2	1	2	2	1	2	1	1
7	2	1	2	2	1	1	2	2	1	2	1
8	2	1	2	1	2	2	2	1	1	1	2
9	2	1	1	2	2	2	1	2	2	1	1
10	2	2	2	1	1	1	1	2	2	1	2
11	2	2	1	2	1	2	1	1	1	2	2
12	2	2	1	1	2	1	2	1	2	2	1

Note: The interaction between any two columns is confounded partially with the remaining nine columns. Do not use this array if the interactions must be estimated.

$L_{16} (2^{15})$

$L_{16} (2^{15})$ Orthogonal Array

Expt. No.	1	2	3	4	5	6	7	8	9	10	11	12	13	14	15
1	1	1	1	1	1	1	1	1	1	1	1	1	1	1	1
2	1	1	1	1	1	1	1	2	2	2	2	2	2	2	2
3	1	1	1	2	2	2	2	1	1	1	1	2	2	2	2
4	1	1	1	2	2	2	2	2	2	2	2	1	1	1	1
5	1	2	2	1	1	2	2	1	1	2	2	1	1	2	2
6	1	2	2	1	1	2	2	2	2	1	1	2	2	1	1
7	1	2	2	2	2	1	1	1	1	2	2	2	2	1	1
8	1	2	2	2	2	1	1	2	2	1	1	1	1	2	2
9	2	1	2	1	2	1	2	1	2	1	2	1	2	1	2
10	2	1	2	1	2	1	2	2	1	2	1	2	1	2	1
11	2	1	2	2	1	2	1	1	2	1	2	2	1	2	1
12	2	1	2	2	1	2	1	2	1	2	1	1	2	1	2
13	2	2	1	1	2	2	1	1	2	2	1	1	2	2	1
14	2	2	1	1	2	2	1	2	1	1	2	2	1	1	2
15	2	2	1	2	1	1	2	1	2	2	1	2	1	1	2
16	2	2	1	2	1	1	2	2	1	1	2	1	2	2	1

Interaction Table for $L_{16} (2^{15})$

Column	1	2	3	4	5	6	7	8	9	10	11	12	13	14	15
1	(1)	3	2	5	4	7	6	9	8	11	10	13	12	15	14
2		(2)	1	6	7	4	5	10	11	8	9	14	15	12	13
3			(3)	7	6	5	4	11	10	9	8	15	14	13	12
4				(4)	1	2	3	12	13	14	15	8	9	10	11
5					(5)	3	2	13	12	15	14	9	8	11	10
6						(6)	1	14	15	12	13	10	11	8	9
7							(7)	15	14	13	12	11	10	9	8
8								(8)	1	2	3	4	5	6	7
9									(9)	3	2	5	4	7	6
10										(10)	1	6	7	4	5
11											(11)	7	6	5	4
12												(12)	1	2	3
13													(13)	3	2
14														(14)	1
15															(15)

Linear Graphs for L_{16}

(1)

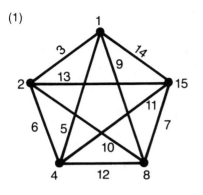

(2)

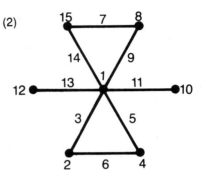

(3)

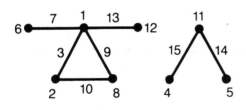

(4)

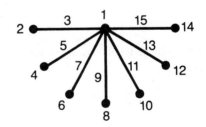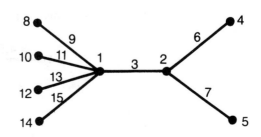

(5)

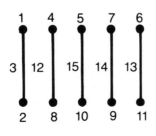

(6)

L'_{16} (4^5) Orthogonal Array

Expt. No.	Column				
	1	2	3	4	5
1	1	1	1	1	1
2	1	2	2	2	2
3	1	3	3	3	3
4	1	4	4	4	4
5	2	1	2	3	4
6	2	2	1	4	3
7	2	3	4	1	2
8	2	4	3	2	1
9	3	1	3	4	2
10	3	2	4	3	1
11	3	3	1	2	4
12	3	4	2	1	3
13	4	1	4	2	3
14	4	2	3	1	4
15	4	3	2	4	1
16	4	4	1	3	2

Note: To estimate the interaction between columns 1 and 2, all other columns must be kept empty.

Linear Graph for L'_{16}

$$L_{18} \ (2^1 \times 3^7)$$

$L_{18} \ (2^1 \times 3^7)$ Orthogonal Array

Expt. No.	Column							
	1	2	3	4	5	6	7	8
1	1	1	1	1	1	1	1	1
2	1	1	2	2	2	2	2	2
3	1	1	3	3	3	3	3	3
4	1	2	1	1	2	2	3	3
5	1	2	2	2	3	3	1	1
6	1	2	3	3	1	1	2	2
7	1	3	1	2	1	3	2	3
8	1	3	2	3	2	1	3	1
9	1	3	3	1	3	2	1	2
10	2	1	1	3	3	2	2	1
11	2	1	2	1	1	3	3	2
12	2	1	3	2	2	1	1	3
13	2	2	1	2	3	1	3	2
14	2	2	2	3	1	2	1	3
15	2	2	3	1	2	3	2	1
16	2	3	1	3	2	3	1	2
17	2	3	2	1	3	1	2	3
18	2	3	3	2	1	2	3	1

Note: Interaction between columns 1 and 2 is orthogonal to all columns and hence can be estimated without sacrificing any column. The interaction can be estimated from the 2-way table of columns 1 and 2. Columns 1 and 2 can be combined to form a 6-level column. Interactions between any other pair of columns is confounded partially with the remaining columns.

Linear Graph for L_{18}

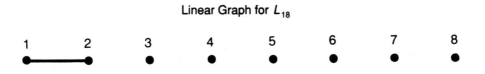

L_{25} (5^6) Orthogonal Array

Expt. No.	Column 1	2	3	4	5	6
1	1	1	1	1	1	1
2	1	2	2	2	2	2
3	1	3	3	3	3	3
4	1	4	4	4	4	4
5	1	5	5	5	5	5
6	2	1	2	3	4	5
7	2	2	3	4	5	1
8	2	3	4	5	1	2
9	2	4	5	1	2	3
10	2	5	1	2	3	4
11	3	1	3	5	2	4
12	3	2	4	1	3	5
13	3	3	5	2	4	1
14	3	4	1	3	5	2
15	3	5	2	4	1	3
16	4	1	4	2	5	3
17	4	2	5	3	1	4
18	4	3	1	4	2	5
19	4	4	2	5	3	1
20	4	5	3	1	4	2
21	5	1	5	4	3	2
22	5	2	1	5	4	3
23	5	3	2	1	5	4
24	5	4	3	2	1	5
25	5	5	4	3	2	1

Note: To estimate the interaction between columns 1 and 2, all other columns must be kept empty.

Linear Graph for L_{25}

3,4,5,6

1 ●━━━━━━● 2

L_{27} (3^{13}) Orthogonal Array

Expt. No.	Column												
	1	2	3	4	5	6	7	8	9	10	11	12	13
1	1	1	1	1	1	1	1	1	1	1	1	1	1
2	1	1	1	1	2	2	2	2	2	2	2	2	2
3	1	1	1	1	3	3	3	3	3	3	3	3	3
4	1	2	2	2	1	1	1	2	2	2	3	3	3
5	1	2	2	2	2	2	2	3	3	3	1	1	1
6	1	2	2	2	3	3	3	1	1	1	2	2	2
7	1	3	3	3	1	1	1	3	3	3	2	2	2
8	1	3	3	3	2	2	2	1	1	1	3	3	3
9	1	3	3	3	3	3	3	2	2	2	1	1	1
10	2	1	2	3	1	2	3	1	2	3	1	2	3
11	2	1	2	3	2	3	1	2	3	1	2	3	1
12	2	1	2	3	3	1	2	3	1	2	3	1	2
13	2	2	3	1	1	2	3	2	3	1	3	1	2
14	2	2	3	1	2	3	1	3	1	2	1	2	3
15	2	2	3	1	3	1	2	1	2	3	2	3	1
16	2	3	1	2	1	2	3	3	1	2	2	3	1
17	2	3	1	2	2	3	1	1	2	3	3	1	2
18	2	3	1	2	3	1	2	2	3	1	1	2	3
19	3	1	3	2	1	3	2	1	3	2	1	3	2
20	3	1	3	2	2	1	3	2	1	3	2	1	3
21	3	1	3	2	3	2	1	3	2	1	3	2	1
22	3	2	1	3	1	3	2	2	1	3	3	2	1
23	3	2	1	3	2	1	3	3	2	1	1	3	2
24	3	2	1	3	3	2	1	1	3	2	2	1	3
25	3	3	2	1	1	3	2	3	2	1	2	1	3
26	3	3	2	1	2	1	3	1	3	2	3	2	1
27	3	3	2	1	3	2	1	2	1	3	1	3	2

Interaction Table for L_{27} (3^{13})

Column	1	2	3	4	5	6	7	8	9	10	11	12	13
1	(1)	3 4	2 4	2 3	6 7	5 7	5 6	9 10	8 10	8 9	12 13	11 13	11 12
2		(2)	1 4	1 3	8 11	9 12	10 13	5 11	6 12	7 13	5 8	6 9	7 10
3			(3)	1 2	9 13	10 11	8 12	7 12	5 13	6 11	6 10	7 8	5 9
4				(4)	10 12	8 13	9 11	6 13	7 11	5 12	7 9	5 10	6 8
5					(5)	1 7	1 6	2 11	3 13	4 12	2 8	4 10	3 9
6						(6)	1 5	4 13	2 12	3 11	3 10	2 9	4 8
7							(7)	3 12	4 11	2 13	4 9	3 8	2 10
8								(8)	1 10	1 9	2 5	3 7	4 6
9									(9)	1 8	4 7	2 6	3 5
10										(10)	3 6	4 5	2 7
11											(11)	1 13	1 12
12												(12)	1 11
13													(13)

Linear Graph for L_{27}

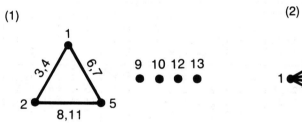

$$L_{32} (2^{31})$$

$L_{32} (2^{31})$ Orthogonal Array

Expt. No.	1	2	3	4	5	6	7	8	9	10	11	12	13	14	15	16	17	18	19	20	21	22	23	24	25	26	27	28	29	30	31
1	1	1	1	1	1	1	1	1	1	1	1	1	1	1	1	1	1	1	1	1	1	1	1	1	1	1	1	1	1	1	1
2	1	1	1	1	1	1	1	1	1	1	1	1	1	1	1	1	2	2	2	2	2	2	2	2	2	2	2	2	2	2	2
3	1	1	1	1	1	1	1	2	2	2	2	2	2	2	2	1	1	1	1	1	1	1	1	2	2	2	2	2	2	2	2
4	1	1	1	1	1	1	1	2	2	2	2	2	2	2	2	2	2	2	2	2	2	2	2	1	1	1	1	1	1	1	1
5	1	1	1	2	2	2	2	1	1	1	1	2	2	2	2	1	1	1	1	2	2	2	2	1	1	1	1	2	2	2	2
6	1	1	1	2	2	2	2	1	1	1	1	2	2	2	2	2	2	2	2	1	1	1	1	2	2	2	2	1	1	1	1
7	1	1	1	2	2	2	2	2	2	2	2	1	1	1	1	1	1	1	1	2	2	2	2	2	2	2	2	1	1	1	1
8	1	1	1	2	2	2	2	2	2	2	2	1	1	1	1	2	2	2	2	1	1	1	1	1	1	1	1	2	2	2	2
9	1	2	2	1	1	2	2	1	1	2	2	1	1	2	2	1	1	2	2	1	1	2	2	1	1	2	2	1	1	2	2
10	1	2	2	1	1	2	2	1	1	2	2	1	1	2	2	2	2	1	1	2	2	1	1	2	2	1	1	2	2	1	1
11	1	2	2	1	1	2	2	2	2	1	1	2	2	1	1	1	1	2	2	1	1	2	2	2	2	1	1	2	2	1	1
12	1	2	2	1	1	2	2	2	2	1	1	2	2	1	1	2	2	1	1	2	2	1	1	1	1	2	2	1	1	2	2
13	1	2	2	2	2	1	1	1	1	2	2	2	2	1	1	1	1	2	2	2	2	1	1	1	1	2	2	2	2	1	1
14	1	2	2	2	2	1	1	1	1	2	2	2	2	1	1	2	2	1	1	1	1	2	2	2	2	1	1	1	1	2	2
15	1	2	2	2	2	1	1	2	2	1	1	1	1	2	2	1	1	2	2	2	2	1	1	2	2	1	1	1	1	2	2
16	1	2	2	2	2	1	1	2	2	1	1	1	1	2	2	2	2	1	1	1	1	2	2	1	1	2	2	2	2	1	1
17	2	1	2	1	2	1	2	1	2	1	2	1	2	1	2	1	2	1	2	1	2	1	2	1	2	1	2	1	2	1	2
18	2	1	2	1	2	1	2	1	2	1	2	1	2	1	2	2	1	2	1	2	1	2	1	2	1	2	1	2	1	2	1
19	2	1	2	1	2	1	2	2	1	2	1	2	1	2	1	1	2	1	2	1	2	1	2	2	1	2	1	2	1	2	1
20	2	1	2	1	2	1	2	2	1	2	1	2	1	2	1	2	1	2	1	2	1	2	1	1	2	1	2	1	2	1	2
21	2	1	2	2	1	2	1	1	2	1	2	2	1	2	1	1	2	1	2	2	1	2	1	1	2	1	2	2	1	2	1
22	2	1	2	2	1	2	1	1	2	1	2	2	1	2	1	2	1	2	1	1	2	1	2	2	1	2	1	1	2	1	2
23	2	1	2	2	1	2	1	2	1	2	1	1	2	1	2	1	2	1	2	2	1	2	1	2	1	2	1	1	2	1	2
24	2	1	2	2	1	2	1	2	1	2	1	1	2	1	2	2	1	2	1	1	2	1	2	1	2	1	2	2	1	2	1
25	2	2	1	1	2	2	1	1	2	2	1	1	2	2	1	1	2	2	1	1	2	2	1	1	2	2	1	1	2	2	1
26	2	2	1	1	2	2	1	1	2	2	1	1	2	2	1	2	1	1	2	2	1	1	2	2	1	1	2	2	1	1	2
27	2	2	1	1	2	2	1	2	1	1	2	2	1	1	2	1	2	2	1	1	2	2	1	2	1	1	2	2	1	1	2
28	2	2	1	1	2	2	1	2	1	1	2	2	1	1	2	2	1	1	2	2	1	1	2	1	2	2	1	1	2	2	1
29	2	2	1	2	1	1	2	1	2	2	1	2	1	1	2	1	2	2	1	2	1	1	2	1	2	2	1	2	1	1	2
30	2	2	1	2	1	1	2	1	2	2	1	2	1	1	2	2	1	1	2	1	2	2	1	2	1	1	2	1	2	2	1
31	2	2	1	2	1	1	2	2	1	1	2	1	2	2	1	1	2	2	1	2	1	1	2	2	1	1	2	1	2	2	1
32	2	2	1	2	1	1	2	2	1	1	2	1	2	2	1	2	1	1	2	1	2	2	1	1	2	2	1	2	1	1	2

Interaction Table for L_{32} (2^{31})

Column	1	2	3	4	5	6	7	8	9	10	11	12	13	14	15	16	17	18	19	20	21	22	23	24	25	26	27	28	29	30	31
1	(1)	3	2	5	4	7	6	9	8	11	10	13	12	15	14	17	16	19	18	21	20	23	22	25	24	27	26	29	28	31	30
2		(2)	1	6	7	4	5	10	11	8	9	14	15	12	13	18	19	16	17	22	23	20	21	26	27	24	25	30	31	28	29
3			(3)	7	6	5	4	11	10	9	8	15	14	13	12	19	18	17	16	23	22	21	20	27	26	25	24	31	30	29	28
4				(4)	1	2	3	12	13	14	15	8	9	10	11	20	21	22	23	16	17	18	19	28	29	30	31	24	25	26	27
5					(5)	3	2	13	12	15	14	9	8	11	10	21	20	23	22	17	16	19	18	29	28	31	30	25	24	27	26
6						(6)	1	14	15	12	13	10	11	8	9	22	23	20	21	18	19	16	17	30	31	28	29	26	27	24	25
7							(7)	15	14	13	12	11	10	9	8	23	22	21	20	19	18	17	16	31	30	29	28	27	26	25	24
8								(8)	1	2	3	4	5	6	7	24	25	26	27	28	29	30	31	16	17	18	19	20	21	22	23
9									(9)	3	2	5	4	7	6	25	24	27	26	29	28	31	30	17	16	19	18	21	20	23	22
10										(10)	1	6	7	4	5	26	27	24	25	30	31	28	29	18	19	16	17	22	23	20	21
11											(11)	7	6	5	4	27	26	25	24	31	30	29	28	19	18	17	16	23	22	21	20
12												(12)	1	2	3	28	29	30	31	24	25	26	27	20	21	22	23	16	17	18	19
13													(13)	3	2	29	28	31	30	25	24	27	26	21	20	23	22	17	16	19	18
14														(14)	1	30	31	28	29	26	27	24	25	22	23	20	21	18	19	16	17
15															(15)	31	30	29	28	27	26	25	24	23	22	21	20	19	18	17	16
16																(16)	1	2	3	4	5	6	7	8	9	10	11	12	13	14	15
17																	(17)	3	2	5	4	7	6	9	8	11	10	13	12	15	14
18																		(18)	1	6	7	4	5	10	11	8	9	14	15	12	13
19																			(19)	7	6	5	4	11	10	9	8	15	14	13	12
20																				(20)	1	2	3	12	13	14	15	8	9	10	11
21																					(21)	3	2	13	12	15	14	9	8	11	10
22																						(22)	1	14	15	12	13	10	11	8	9
23																							(23)	15	14	13	12	11	10	9	8
24																								(24)	1	2	3	4	5	6	7
25																									(25)	3	2	5	4	7	6
26																										(26)	1	6	7	4	5
27																											(27)	7	6	5	4
28																												(28)	1	2	3
29																													(29)	3	2
30																														(30)	1
31																															(31)

Linear Graphs for L_{32}

(1)

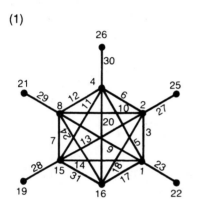

(2)

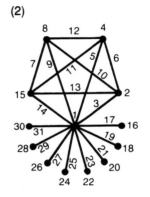

(3)

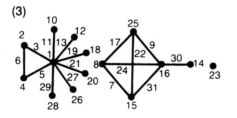

(4)

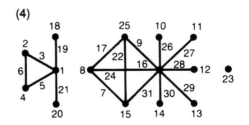

Linear Graphs for L_{32} (Continued)

(5)

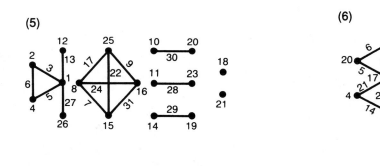

(6)

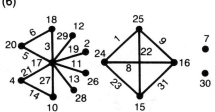

(7)

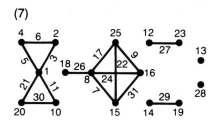

(8)

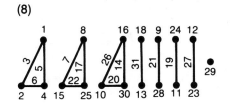

Linear Graphs for L_{32} (Continued)

(9)

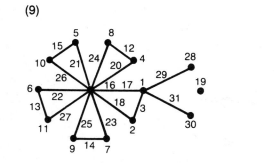

(10)

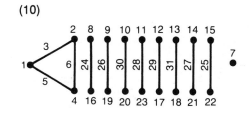

(11)

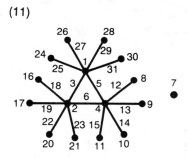

(12)

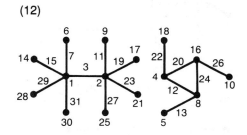

(13)

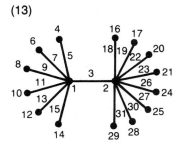

$$L'_{32} \ (2^1 \times 4^9)$$

$L'_{32} \ (2^1 \times 4^9)$ Orthogonal Array

Expt. No.	Column									
	1	2	3	4	5	6	7	8	9	10
1	1	1	1	1	1	1	1	1	1	1
2	1	1	2	2	2	2	2	2	2	2
3	1	1	3	3	3	3	3	3	3	3
4	1	1	4	4	4	4	4	4	4	4
5	1	2	1	1	2	2	3	3	4	4
6	1	2	2	2	1	1	4	4	3	3
7	1	2	3	3	4	4	1	1	2	2
8	1	2	4	4	3	3	2	2	1	1
9	1	3	1	2	3	4	1	2	3	4
10	1	3	2	1	4	3	2	1	4	3
11	1	3	3	4	1	2	3	4	1	2
12	1	3	4	3	2	1	4	3	2	1
13	1	4	1	2	4	3	3	4	2	1
14	1	4	2	1	3	4	4	3	1	2
15	1	4	3	4	2	1	1	2	4	3
16	1	4	4	3	1	2	2	1	3	4
17	2	1	1	4	1	4	2	3	2	3
18	2	1	2	3	2	3	1	4	1	4
19	2	1	3	2	3	2	4	1	4	1
20	2	1	4	1	4	1	3	2	3	2
21	2	2	1	4	2	3	4	1	3	2
22	2	2	2	3	1	4	3	2	4	1
23	2	2	3	2	4	1	2	3	1	4
24	2	2	4	1	3	2	1	4	2	3
25	2	3	1	3	3	1	2	4	4	2
26	2	3	2	4	4	2	1	3	3	1
27	2	3	3	1	1	3	4	2	2	4
28	2	3	4	2	2	4	3	1	1	3
29	2	4	1	3	4	2	4	2	1	3
30	2	4	2	4	3	1	3	1	2	4
31	2	4	3	1	2	4	2	4	3	1
32	2	4	4	2	1	3	1	3	4	2

Note: Interaction between columns 1 and 2 is orthogonal to all columns and hence can be estimated without sacrificing any column. It can be estimated from the 2-way table of these columns. Columns 1 and 2 can be combined to form an 8-level column. Interactions between any two 4-level columns is confounded partially with each of the remaining 4-level columns.

Linear Graph for L'_{32}

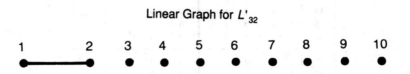

$$\boxed{L_{36} \ (2^{11} \times 3^{12})}$$

$L_{36} \ (2^{11} \times 3^{12})$ Orthogonal Array

Expt. No.	Column																									
	1	2	3	4	5	6	7	8	9	10	11		12	13	14	15	16	17	18	19	20	21	22	23
1	1	1	1	1	1	1	1	1	1	1	1		1	1	1	1	1	1	1	1	1	1	1	1
2	1	1	1	1	1	1	1	1	1	1	1		2	2	2	2	2	2	2	2	2	2	2	2
3	1	1	1	1	1	1	1	1	1	1	1		3	3	3	3	3	3	3	3	3	3	3	3
4	1	1	1	1	1	2	2	2	2	2	2		1	1	1	1	2	2	2	2	3	3	3	3
5	1	1	1	1	1	2	2	2	2	2	2		2	2	2	2	3	3	3	3	1	1	1	1
6	1	1	1	1	1	2	2	2	2	2	2		3	3	3	3	1	1	1	1	2	2	2	2
7	1	1	2	2	2	1	1	1	2	2	2		1	1	2	3	1	2	3	3	1	2	2	3
8	1	1	2	2	2	1	1	1	2	2	2		2	2	3	1	2	3	1	1	2	3	3	1
9	1	1	2	2	2	1	1	1	2	2	2		3	3	1	2	3	1	2	2	3	1	1	2
10	1	2	1	2	2	1	2	2	1	1	2		1	1	3	2	1	3	2	3	2	1	3	2
11	1	2	1	2	2	1	2	2	1	1	2		2	2	1	3	2	1	3	1	3	2	1	3
12	1	2	1	2	2	1	2	2	1	1	2		3	3	2	1	3	2	1	2	1	3	2	1
13	1	2	2	1	2	2	1	2	1	2	1		1	2	3	1	3	2	1	3	3	2	1	2
14	1	2	2	1	2	2	1	2	1	2	1		2	3	1	2	1	3	2	1	1	3	2	3
15	1	2	2	1	2	2	1	2	1	2	1		3	1	2	3	2	1	3	2	2	1	3	1
16	1	2	2	2	1	2	2	1	2	1	1		1	2	3	2	1	1	3	2	3	3	2	1
17	1	2	2	2	1	2	2	1	2	1	1		2	3	1	3	2	2	1	3	1	1	3	2
18	1	2	2	2	1	2	2	1	2	1	1		3	1	2	1	3	3	2	1	2	2	1	3
19	2	1	2	2	1	1	2	2	1	2	1		1	2	1	3	3	3	1	2	2	1	2	3
20	2	1	2	2	1	1	2	2	1	2	1		2	3	2	1	1	1	2	3	3	2	3	1
21	2	1	2	2	1	1	2	2	1	2	1		3	1	3	2	2	2	3	1	1	3	1	2
22	2	1	2	1	2	2	2	1	1	1	2		1	2	2	3	3	1	2	1	1	3	3	2
23	2	1	2	1	2	2	2	1	1	1	2		2	3	3	1	1	2	3	2	2	1	1	3
24	2	1	2	1	2	2	2	1	1	1	2		3	1	1	2	2	3	1	3	3	2	2	1
25	2	1	1	2	2	2	1	2	2	1	1		1	3	2	1	2	3	3	1	3	1	2	2
26	2	1	1	2	2	2	1	2	2	1	1		2	1	3	2	3	1	1	2	1	2	3	3
27	2	1	1	2	2	2	1	2	2	1	1		3	2	1	3	1	2	2	3	2	3	1	1
28	2	2	2	1	1	1	1	1	2	2	1	2	1	3	2	2	2	1	1	3	2	3	1	3
29	2	2	2	1	1	1	1	1	2	2	1	2	2	1	3	3	3	2	2	1	3	1	2	1
30	2	2	2	1	1	1	1	1	2	2	1	2	3	2	1	1	1	3	3	2	1	2	3	2
31	2	2	1	2	1	2	1	1	1	2	2		1	3	3	3	2	3	2	2	1	2	1	1
32	2	2	1	2	1	2	1	1	1	2	2		2	1	1	1	3	1	3	3	2	3	2	2
33	2	2	1	2	1	2	1	1	1	2	2		3	2	2	2	1	2	1	1	3	1	3	3
34	2	2	1	1	2	1	2	1	2	2	1		1	3	1	2	3	2	3	1	2	2	3	1
35	2	2	1	1	2	1	2	1	2	2	1		2	1	2	3	1	3	1	2	3	3	1	2
36	2	2	1	1	2	1	2	1	2	2	1		3	2	3	1	2	1	2	3	1	1	2	3

Note: Interaction between any two columns is partially confounded with the remaining columns.

$$L'_{36} \ (2^3 \times 3^{13})$$

$L'_{36} \ (2^3 \times 3^{13})$ Orthogonal Array

Expt. No.	Column 1	2	3	4	5	6	7	8	9	10	11	12	13	14	15	16
1	1	1	1	1	1	1	1	1	1	1	1	1	1	1	1	1
2	1	1	1	1	2	2	2	2	2	2	2	2	2	2	2	2
3	1	1	1	1	3	3	3	3	3	3	3	3	3	3	3	3
4	1	2	2	1	1	1	1	1	2	2	2	2	3	3	3	3
5	1	2	2	1	2	2	2	2	3	3	3	3	1	1	1	1
6	1	2	2	1	3	3	3	3	1	1	1	1	2	2	2	2
7	2	1	2	1	1	1	2	3	1	2	3	3	1	2	2	3
8	2	1	2	1	2	2	3	1	2	3	1	1	2	3	3	1
9	2	1	2	1	3	3	1	2	3	1	2	2	3	1	1	2
10	2	2	1	1	1	1	3	2	1	3	2	3	2	1	3	2
11	2	2	1	1	2	2	1	3	2	1	3	1	3	2	1	3
12	2	2	1	1	3	3	2	1	3	2	1	2	1	3	2	1
13	1	1	1	2	1	2	3	1	3	2	1	3	3	2	1	2
14	1	1	1	2	2	3	1	2	1	3	2	1	1	3	2	3
15	1	1	1	2	3	1	2	3	2	1	3	2	2	1	3	1
16	1	2	2	2	1	2	3	2	1	1	3	2	3	3	2	1
17	1	2	2	2	2	3	1	3	2	2	1	3	1	1	3	2
18	1	2	2	2	3	1	2	1	3	3	2	1	2	2	1	3
19	2	1	2	2	1	2	1	3	3	3	1	2	2	1	2	3
20	2	1	2	2	2	3	2	1	1	1	2	3	3	2	3	1
21	2	1	2	2	3	1	3	2	2	2	3	1	1	3	1	2
22	2	2	1	2	1	2	2	3	3	1	2	1	1	3	3	2
23	2	2	1	2	2	3	3	1	1	2	3	2	2	1	1	3
24	2	2	1	2	3	1	1	2	2	3	1	3	3	2	2	1
25	1	1	1	3	1	3	2	1	2	3	3	1	3	1	2	2
26	1	1	1	3	2	1	3	2	3	1	1	2	1	2	3	3
27	1	1	1	3	3	2	1	3	1	2	2	3	2	3	1	1
28	1	2	2	3	1	3	2	2	2	1	1	3	2	3	1	3
29	1	2	2	3	2	1	3	3	3	2	2	1	3	1	2	1
30	1	2	2	3	3	2	1	1	1	3	3	2	1	2	3	2
31	2	1	2	3	1	3	3	3	2	3	2	2	1	2	1	1
32	2	1	2	3	2	1	1	1	3	1	3	3	2	3	2	2
33	2	1	2	3	3	2	2	2	1	2	1	1	3	1	3	3
34	2	2	1	3	1	3	1	2	3	2	3	1	2	2	2	3
35	2	2	1	3	2	1	2	3	1	3	1	2	3	3	1	2
36	2	2	1	3	3	2	3	1	2	1	2	3	1	1	2	3

Notes: (i) The interactions 1×4, 2×4 and 3×4 are orthogonal to all columns and hence can be obtained without sacrificing any column. (ii) The 3-factor interaction between columns 1, 2, and 4 can be obtained by keeping only column 3 empty. Thus, a 12-level factor can be formed by combining columns 1, 2, and 4 and by keeping column 3 empty. (iii) Columns 5 through 16 in the array $L'_{36}(2^3 \times 3^{13})$ are the same as the columns 12 through 23 in the array $L_{36}(2^{11} \times 3^{12})$.

Linear Graphs for L'_{36}

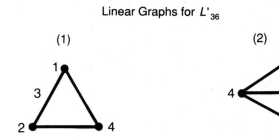

$$L_{50} \ (2^1 \times 5^{11})$$

$L_{50} \ (2^1 \times 5^{11})$ Orthogonal Array

Expt. No.	Column 1	2	3	4	5	6	7	8	9	10	11	12
1	1	1	1	1	1	1	1	1	1	1	1	1
2	1	1	2	2	2	2	2	2	2	2	2	2
3	1	1	3	3	3	3	3	3	3	3	3	3
4	1	1	4	4	4	4	4	4	4	4	4	4
5	1	1	5	5	5	5	5	5	5	5	5	5
6	1	2	1	2	3	4	5	1	2	3	4	5
7	1	2	2	3	4	5	1	2	3	4	5	1
8	1	2	3	4	5	1	2	3	4	5	1	2
9	1	2	4	5	1	2	3	4	5	1	2	3
10	1	2	5	1	2	3	4	5	1	2	3	4
11	1	3	1	3	5	2	4	4	1	3	5	2
12	1	3	2	4	1	3	5	5	2	4	1	3
13	1	3	3	5	2	4	1	1	3	5	2	4
14	1	3	4	1	3	5	2	2	4	1	3	5
15	1	3	5	2	4	1	3	3	5	2	4	1
16	1	4	1	4	2	5	3	5	3	1	4	2
17	1	4	2	5	3	1	4	1	4	2	5	3
18	1	4	3	1	4	2	5	2	5	3	1	4
19	1	4	4	2	5	3	1	3	1	4	2	5
20	1	4	5	3	1	4	2	4	2	5	3	1
21	1	5	1	5	4	3	2	4	3	2	1	5
22	1	5	2	1	5	4	3	5	4	3	2	1
23	1	5	3	2	1	5	4	1	5	4	3	2
24	1	5	4	3	2	1	5	2	1	5	4	3
25	1	5	5	4	3	2	1	3	2	1	5	4
26	2	1	1	1	4	5	4	3	2	5	2	3
27	2	1	2	2	5	1	5	4	3	1	3	4
28	2	1	3	3	1	2	1	5	4	2	4	5
29	2	1	4	4	2	3	2	1	5	3	5	1
30	2	1	5	5	3	4	3	2	1	4	1	2
31	2	2	1	2	1	3	3	2	4	5	5	4
32	2	2	2	3	2	4	4	3	5	1	1	5
33	2	2	3	4	3	5	5	4	1	2	2	1
34	2	2	4	5	4	1	1	5	2	3	3	2
35	2	2	5	1	5	2	2	1	3	4	4	3
36	2	3	1	3	3	1	2	5	5	4	2	4
37	2	3	2	4	4	2	3	1	1	5	3	5
38	2	3	3	5	5	3	4	2	2	1	4	1
39	2	3	4	1	1	4	5	3	3	2	5	2
40	2	3	5	2	2	5	1	4	4	3	1	3

(Continued)

L_{50} ($2^1 \times 5^{11}$) (Continued)

Expt. No.	Column											
	1	2	3	4	5	6	7	8	9	10	11	12
41	2	4	1	4	5	4	1	2	5	2	3	3
42	2	4	2	5	1	5	2	3	1	3	4	4
43	2	4	3	1	2	1	3	4	2	4	5	5
44	2	4	4	2	3	2	4	5	3	5	1	1
45	2	4	5	3	4	3	5	1	4	1	2	2
46	2	5	1	5	2	2	5	3	4	4	3	1
47	2	5	2	1	3	3	1	4	5	5	4	2
48	2	5	3	2	4	4	2	5	1	1	5	3
49	2	5	4	3	5	5	3	1	2	2	1	4
50	2	5	5	4	1	1	4	2	3	3	2	5

Note: Interaction between columns 1 and 2 is orthogonal to all columns and hence can be estimated without sacrificing any column. It can be estimated from the 2-way table of these two columns. Columns 1 and 2 can be combined to form a 10-level column.

Linear Graphs for L_{50}

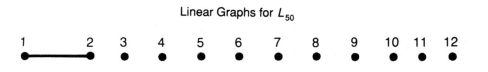

$$L_{54}\ (2^1 \times 3^{25})$$

$L_{54}\ (2^1 \times 3^{25})$ Orthogonal Array

Expt. No.	1	2	3	4	5	6	7	8	9	10	11	12	13	14	15	16	17	18	19	20	21	22	23	24	25	26
1	1	1	1	1	1	1	1	1	1	1	1	1	1	1	1	1	1	1	1	1	1	1	1	1	1	1
2	1	1	1	1	1	1	1	1	1	2	2	2	2	2	2	2	2	2	2	2	2	2	2	2	2	2
3	1	1	1	1	1	1	1	1	1	3	3	3	3	3	3	3	3	3	3	3	3	3	3	3	3	3
4	1	1	2	2	2	2	2	2	1	1	1	1	1	1	2	3	2	3	2	3	2	3	2	3	2	3
5	1	1	2	2	2	2	2	2	2	2	2	2	2	2	3	1	3	1	3	1	3	1	3	1	3	1
6	1	1	2	2	2	2	2	2	3	3	3	3	3	3	1	2	1	2	1	2	1	2	1	2	1	2
7	1	1	3	3	3	3	3	3	1	1	1	1	1	1	3	2	3	2	3	2	3	2	3	2	3	2
8	1	1	3	3	3	3	3	3	2	2	2	2	2	2	1	3	1	3	1	3	1	3	1	3	1	3
9	1	1	3	3	3	3	3	3	3	3	3	3	3	3	2	1	2	1	2	1	2	1	2	1	2	1
10	1	2	1	1	2	2	3	3	1	1	2	2	3	3	1	1	1	1	2	3	2	3	3	2	3	2
11	1	2	1	1	2	2	3	3	2	2	3	3	1	1	2	2	2	2	3	1	3	1	1	3	1	3
12	1	2	1	1	2	2	3	3	3	3	1	1	2	2	3	3	3	3	1	2	1	2	2	1	2	1
13	1	2	2	2	3	3	1	1	1	1	2	2	3	3	2	3	2	3	3	2	3	2	1	1	1	1
14	1	2	2	2	3	3	1	1	2	2	3	3	1	1	3	1	3	1	1	3	1	3	2	2	2	2
15	1	2	2	2	3	3	1	1	3	3	1	1	2	2	1	2	1	2	2	1	2	1	3	3	3	3
16	1	2	3	3	1	1	2	2	1	1	2	2	3	3	3	2	3	2	1	1	1	1	2	3	2	3
17	1	2	3	3	1	1	2	2	2	2	3	3	1	1	1	3	1	3	2	2	2	2	3	1	3	1
18	1	2	3	3	1	1	2	2	3	3	1	1	2	2	2	1	2	1	3	3	3	3	1	2	1	2
19	1	3	1	2	1	3	2	3	1	2	1	3	2	3	1	1	2	3	1	1	3	2	2	3	3	2
20	1	3	1	2	1	3	2	3	2	3	2	1	3	1	2	2	3	1	2	2	1	3	3	1	1	3
21	1	3	1	2	1	3	2	3	3	1	3	2	1	2	3	3	1	2	3	3	2	1	1	2	2	1
22	1	3	2	3	2	1	3	1	1	2	1	3	2	3	2	3	3	2	2	3	1	1	3	2	1	1
23	1	3	2	3	2	1	3	1	2	3	2	1	3	1	3	1	1	3	3	1	2	2	1	3	2	2
24	1	3	2	3	2	1	3	1	3	1	3	2	1	2	1	2	2	1	1	2	3	3	2	1	3	3
25	1	3	3	1	3	2	1	2	1	2	1	3	2	3	3	2	1	1	3	2	2	3	1	1	2	3
26	1	3	3	1	3	2	1	2	2	3	2	1	3	1	1	3	2	2	1	3	3	1	2	2	3	1
27	1	3	3	1	3	2	1	2	3	1	3	2	1	2	2	1	3	3	2	1	1	2	3	3	1	2
28	2	1	1	3	3	2	2	1	1	3	3	2	2	1	1	1	3	2	3	2	2	3	2	3	1	1
29	2	1	1	3	3	2	2	1	2	1	1	3	3	2	2	2	1	3	1	3	3	1	3	1	2	2
30	2	1	1	3	3	2	2	1	3	2	2	1	1	3	3	3	2	1	2	1	1	2	1	2	3	3
31	2	1	2	1	1	3	3	2	1	3	3	2	2	1	2	3	1	1	1	1	3	2	3	2	2	3
32	2	1	2	1	1	3	3	2	2	1	1	3	3	2	3	1	2	2	2	2	1	3	1	3	3	1
33	2	1	2	1	1	3	3	2	3	2	2	1	1	3	1	2	3	3	3	3	2	1	2	1	1	2
34	2	1	3	2	2	1	1	3	1	3	3	2	2	1	3	2	2	3	2	3	1	1	1	1	3	2
35	2	1	3	2	2	1	1	3	2	1	1	3	3	2	1	3	3	1	3	1	2	2	2	2	1	3
36	2	1	3	2	2	1	1	3	3	2	2	1	1	3	2	1	1	2	1	2	3	3	3	3	2	1
37	2	2	1	2	3	1	3	2	1	2	3	1	3	2	1	1	2	3	3	2	1	1	3	2	2	3
38	2	2	1	2	3	1	3	2	2	3	1	2	1	3	2	2	3	1	1	3	2	2	1	3	3	1
39	2	2	1	2	3	1	3	2	3	1	2	3	2	1	3	3	1	2	2	1	3	3	2	1	1	2

(Continued)

$$L_{54} \ (2^1 \times 3^{25}) \ \text{(Continued)}$$

Expt. No.	1	2	3	4	5	6	7	8	9	10	11	12	13	14	15	16	17	18	19	20	21	22	23	24	25	26
40	2	2	2	3	1	2	1	3	1	2	3	1	3	2	2	3	3	2	1	1	2	3	1	1	3	2
41	2	2	2	3	1	2	1	3	2	3	1	2	1	3	3	1	1	3	2	2	3	1	2	2	1	3
42	2	2	2	3	1	2	1	3	3	1	2	3	2	1	1	2	2	1	3	3	1	2	3	3	2	1
43	2	2	3	1	2	3	2	1	1	2	3	1	3	2	3	2	1	1	2	3	3	2	2	3	1	1
44	2	2	3	1	2	3	2	1	2	3	1	2	1	3	1	3	2	2	3	1	1	3	3	1	2	2
45	2	2	3	1	2	3	2	1	3	1	2	3	2	1	2	1	3	3	1	2	2	1	1	2	3	3
46	2	3	1	3	2	3	1	2	1	3	2	3	1	2	1	1	3	2	2	3	3	2	1	1	2	3
47	2	3	1	3	2	3	1	2	2	1	3	1	2	3	2	2	1	3	3	1	1	3	2	2	3	1
48	2	3	1	3	2	3	1	2	3	2	1	2	3	1	3	3	2	1	1	2	2	1	3	3	1	2
49	2	3	2	1	3	1	2	2	1	3	2	3	1	2	2	3	1	1	3	2	1	1	2	3	3	2
50	2	3	2	1	3	1	2	3	2	1	3	1	2	3	3	1	2	2	1	3	2	2	3	1	1	3
51	2	3	2	1	3	1	2	3	3	2	1	2	3	1	1	2	3	3	2	1	3	3	1	2	2	1
52	2	3	3	2	1	2	3	1	1	3	2	3	1	2	3	2	2	3	1	1	2	3	3	2	1	1
53	2	3	3	2	1	2	3	1	2	1	3	1	2	3	1	3	3	1	2	2	3	1	1	3	2	2
54	2	3	3	2	1	2	3	1	3	2	1	2	3	1	2	1	1	2	3	3	1	2	2	1	3	3

Notes: (i) Interaction between columns 1 and 2 is orthogonal to all columns and hence can be estimated without sacrificing any column. Also, these columns can be combined to form a 6-level column. (ii) The interactions 1×9, 2×9, and $1 \times 2 \times 9$ appear comprehensively in the columns 10, 11, 12, 13, and 14. Hence, the aforementioned interactions can be obtained by keeping columns 10 through 14 empty. Also, columns 1, 2, and 9 can be combined to form a 18-level column by keeping columns 10 through 14 empty.

Linear Graph for L_{54}

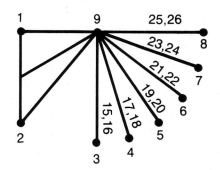

L_{64} (2^{63}) Orthogonal Array

Expt. No.	1	2	3	4	5	6	7	8	9	10	11	12	13	14	15	16	17	18	19	20	21	22	23	24	25	26	27	28	29	30	31
1	1	1	1	1	1	1	1	1	1	1	1	1	1	1	1	1	1	1	1	1	1	1	1	1	1	1	1	1	1	1	1
2	1	1	1	1	1	1	1	1	1	1	1	1	1	1	1	1	1	1	1	1	1	1	1	1	1	1	1	1	1	1	1
3	1	1	1	1	1	1	1	1	1	1	1	1	1	1	1	2	2	2	2	2	2	2	2	2	2	2	2	2	2	2	2
4	1	1	1	1	1	1	1	1	1	1	1	1	1	1	1	2	2	2	2	2	2	2	2	2	2	2	2	2	2	2	2
5	1	1	1	1	1	1	1	2	2	2	2	2	2	2	2	1	1	1	1	1	1	1	1	2	2	2	2	2	2	2	2
6	1	1	1	1	1	1	1	2	2	2	2	2	2	2	2	1	1	1	1	1	1	1	1	2	2	2	2	2	2	2	2
7	1	1	1	1	1	1	1	2	2	2	2	2	2	2	2	2	2	2	2	2	2	2	2	1	1	1	1	1	1	1	1
8	1	1	1	1	1	1	1	2	2	2	2	2	2	2	2	2	2	2	2	2	2	2	2	1	1	1	1	1	1	1	1
9	1	1	1	2	2	2	2	1	1	1	1	2	2	2	2	1	1	1	1	2	2	2	2	1	1	1	1	2	2	2	2
10	1	1	1	2	2	2	2	1	1	1	1	2	2	2	2	1	1	1	1	2	2	2	2	1	1	1	1	2	2	2	2
11	1	1	1	2	2	2	2	1	1	1	1	2	2	2	2	2	2	2	2	1	1	1	1	2	2	2	2	1	1	1	1
12	1	1	1	2	2	2	2	1	1	1	1	2	2	2	2	2	2	2	2	1	1	1	1	2	2	2	2	1	1	1	1
13	1	1	1	2	2	2	2	2	2	2	2	1	1	1	1	1	1	1	1	2	2	2	2	2	2	2	2	1	1	1	1
14	1	1	1	2	2	2	2	2	2	2	2	1	1	1	1	1	1	1	1	2	2	2	2	2	2	2	2	1	1	1	1
15	1	1	1	2	2	2	2	2	2	2	2	1	1	1	1	2	2	2	2	1	1	1	1	1	1	1	1	2	2	2	2
16	1	1	1	2	2	2	2	2	2	2	2	1	1	1	1	2	2	2	2	1	1	1	1	1	1	1	1	2	2	2	2
17	1	2	2	1	1	2	2	1	1	2	2	1	1	2	2	1	1	2	2	1	1	2	2	1	1	2	2	1	1	2	2
18	1	2	2	1	1	2	2	1	1	2	2	1	1	2	2	1	1	2	2	1	1	2	2	1	1	2	2	1	1	2	2
19	1	2	2	1	1	2	2	1	1	2	2	1	1	2	2	2	2	1	1	2	2	1	1	2	2	1	1	2	2	1	1
20	1	2	2	1	1	2	2	1	1	2	2	1	1	2	2	2	2	1	1	2	2	1	1	2	2	1	1	2	2	1	1
21	1	2	2	1	1	2	2	2	2	1	1	2	2	1	1	1	1	2	2	1	1	2	2	2	2	1	1	2	2	1	1
22	1	2	2	1	1	2	2	2	2	1	1	2	2	1	1	1	1	2	2	1	1	2	2	2	2	1	1	2	2	1	1
23	1	2	2	1	1	2	2	2	2	1	1	2	2	1	1	2	2	1	1	2	2	1	1	1	1	2	2	1	1	2	2
24	1	2	2	1	1	2	2	2	2	1	1	2	2	1	1	2	2	1	1	2	2	1	1	1	1	2	2	1	1	2	2
25	1	2	2	2	2	1	1	1	1	2	2	2	2	1	1	1	1	2	2	2	2	1	1	1	1	2	2	2	2	1	1
26	1	2	2	2	2	1	1	1	1	2	2	2	2	1	1	1	1	2	2	2	2	1	1	1	1	2	2	2	2	1	1
27	1	2	2	2	2	1	1	1	1	2	2	2	2	1	1	2	2	1	1	1	1	2	2	2	2	1	1	1	1	2	2
28	1	2	2	2	2	1	1	1	1	2	2	2	2	1	1	2	2	1	1	1	1	2	2	2	2	1	1	1	1	2	2
29	1	2	2	2	2	1	1	2	2	1	1	1	1	2	2	1	1	2	2	2	2	1	1	2	2	1	1	1	1	2	2
30	1	2	2	2	2	1	1	2	2	1	1	1	1	2	2	1	1	2	2	2	2	1	1	2	2	1	1	1	1	2	2
31	1	2	2	2	2	1	1	2	2	1	1	1	1	2	2	2	2	1	1	1	1	2	2	1	1	2	2	2	2	1	1
32	1	2	2	2	2	1	1	2	2	1	1	1	1	2	2	2	2	1	1	1	1	2	2	1	1	2	2	2	2	1	1
33	2	1	2	1	2	1	2	1	2	1	2	1	2	1	2	1	2	1	2	1	2	1	2	1	2	1	2	1	2	1	2
34	2	1	2	1	2	1	2	1	2	1	2	1	2	1	2	1	2	1	2	1	2	1	2	1	2	1	2	1	2	1	2
35	2	1	2	1	2	1	2	1	2	1	2	1	2	1	2	2	1	2	1	2	1	2	1	2	1	2	1	2	1	2	1
36	2	1	2	1	2	1	2	1	2	1	2	1	2	1	2	2	1	2	1	2	1	2	1	2	1	2	1	2	1	2	1
37	2	1	2	1	2	1	2	2	1	2	1	2	1	2	1	1	2	1	2	1	2	1	2	2	1	2	1	2	1	2	1
38	2	1	2	1	2	1	2	2	1	2	1	2	1	2	1	1	2	1	2	1	2	1	2	2	1	2	1	2	1	2	1
39	2	1	2	1	2	1	2	2	1	2	1	2	1	2	1	2	1	2	1	2	1	2	1	1	2	1	2	1	2	1	2
40	2	1	2	1	2	1	2	2	1	2	1	2	1	2	1	2	1	2	1	2	1	2	1	1	2	1	2	1	2	1	2

(Continued)

$$L_{64}\ (2^{63})\ \text{(Continued)}$$

Expt. No.	Column																														
	1	2	3	4	5	6	7	8	9	10	11	12	13	14	15	16	17	18	19	20	21	22	23	24	25	26	27	28	29	30	31
41	2	1	2	2	1	2	1	1	2	1	2	2	1	2	1	1	2	1	2	2	1	2	1	1	2	1	2	2	1	2	1
42	2	1	2	2	1	2	1	1	2	1	2	2	1	2	1	1	2	1	2	2	1	2	1	1	2	1	2	2	1	2	1
43	2	1	2	2	1	2	1	1	2	1	2	2	1	2	1	2	1	2	1	1	2	1	2	2	1	2	1	1	2	1	2
44	2	1	2	2	1	2	1	1	2	1	2	2	1	2	1	2	1	2	1	1	2	1	2	2	1	2	1	1	2	1	2
45	2	1	2	2	1	2	1	2	1	2	1	1	2	1	2	1	2	1	2	2	1	2	1	2	1	2	1	1	2	1	2
46	2	1	2	2	1	2	1	2	1	2	1	1	2	1	2	1	2	1	2	2	1	2	1	2	1	2	1	1	2	1	2
47	2	1	2	2	1	2	1	2	1	2	1	1	2	1	2	2	1	2	1	1	2	1	2	1	2	1	2	2	1	2	1
48	2	1	2	2	1	2	1	2	1	2	1	1	2	1	2	2	1	2	1	1	2	1	2	1	2	1	2	2	1	2	1
49	2	2	1	1	2	2	1	1	2	2	1	1	2	2	1	1	2	2	1	1	2	2	1	1	2	2	1	1	2	2	1
50	2	2	1	1	2	2	1	1	2	2	1	1	2	2	1	1	2	2	1	1	2	2	1	1	2	2	1	1	2	2	1
51	2	2	1	1	2	2	1	1	2	2	1	1	2	2	1	2	1	1	2	2	1	1	2	2	1	1	2	2	1	1	2
52	2	2	1	1	2	2	1	1	2	2	1	1	2	2	1	2	1	1	2	2	1	1	2	2	1	1	2	2	1	1	2
53	2	2	1	1	2	2	1	2	1	1	2	2	1	1	2	1	2	2	1	1	2	2	1	2	1	1	2	2	1	1	2
54	2	2	1	1	2	2	1	2	1	1	2	2	1	1	2	1	2	2	1	1	2	2	1	2	1	1	2	2	1	1	2
55	2	2	1	1	2	2	1	2	1	1	2	2	1	1	2	2	1	1	2	1	2	1	2	2	1	1	2	2	1	1	2
56	2	2	1	1	2	2	1	2	1	1	2	2	1	1	2	2	1	1	2	1	2	1	2	2	1	1	2	2	1	1	2
57	2	2	1	2	1	1	2	1	2	2	1	2	1	1	2	1	2	2	1	2	1	1	2	1	2	2	1	2	1	1	2
58	2	2	1	2	1	1	2	1	2	2	1	2	1	1	2	1	2	2	1	2	1	1	2	1	2	2	1	2	1	1	2
59	2	2	1	2	1	1	2	1	2	2	1	2	1	1	2	2	1	1	2	1	2	2	1	2	1	1	2	1	2	2	1
60	2	2	1	2	1	1	2	1	2	2	1	2	1	1	2	2	1	1	2	1	2	2	1	2	1	1	2	1	2	2	1
61	2	2	1	2	1	1	2	2	1	1	2	1	2	2	1	1	2	2	1	2	1	1	2	2	1	1	2	1	2	2	1
62	2	2	1	2	1	1	2	2	1	1	2	1	2	2	1	1	2	2	1	2	1	1	2	2	1	1	2	1	2	2	1
63	2	2	1	2	1	1	2	2	1	1	2	1	2	2	1	2	1	1	2	1	2	2	1	1	2	2	1	2	1	1	2
64	2	2	1	2	1	1	2	2	1	1	2	1	2	2	1	2	1	1	2	1	2	2	1	1	2	2	1	2	1	1	2

(Continued)

$$L_{64}\ (2^{63})\ \text{(Continued)}$$

Expt. No.	32	33	34	35	36	37	38	39	40	41	42	43	44	45	46	47	48	49	50	51	52	53	54	55	56	57	58	59	60	61	62	63
1	1	1	1	1	1	1	1	1	1	1	1	1	1	1	1	1	1	1	1	1	1	1	1	1	1	1	1	1	1	1	1	1
2	2	2	2	2	2	2	2	2	2	2	2	2	2	2	2	2	2	2	2	2	2	2	2	2	2	2	2	2	2	2	2	2
3	1	1	1	1	1	1	1	1	1	1	1	1	1	1	1	1	2	2	2	2	2	2	2	2	2	2	2	2	2	2	2	2
4	2	2	2	2	2	2	2	2	2	2	2	2	2	2	2	2	1	1	1	1	1	1	1	1	1	1	1	1	1	1	1	1
5	1	1	1	1	1	1	1	1	2	2	2	2	2	2	2	2	1	1	1	1	1	1	1	1	2	2	2	2	2	2	2	2
6	2	2	2	2	2	2	2	2	1	1	1	1	1	1	1	1	2	2	2	2	2	2	2	2	1	1	1	1	1	1	1	1
7	1	1	1	1	1	1	1	1	2	2	2	2	2	2	2	2	2	2	2	2	2	2	2	2	1	1	1	1	1	1	1	1
8	2	2	2	2	2	2	2	2	1	1	1	1	1	1	1	1	1	1	1	1	1	1	1	1	2	2	2	2	2	2	2	2
9	1	1	1	1	2	2	2	2	1	1	1	1	2	2	2	2	1	1	1	1	2	2	2	2	1	1	1	1	2	2	2	2
10	2	2	2	2	1	1	1	1	2	2	2	2	1	1	1	1	2	2	2	2	1	1	1	1	2	2	2	2	1	1	1	1
11	1	1	1	1	2	2	2	2	1	1	1	1	2	2	2	2	2	2	2	2	1	1	1	1	2	2	2	2	1	1	1	1
12	2	2	2	2	1	1	1	1	2	2	2	2	1	1	1	1	1	1	1	1	2	2	2	2	1	1	1	1	2	2	2	2
13	1	1	1	1	2	2	2	2	2	2	2	2	1	1	1	1	1	1	1	1	2	2	2	2	2	2	2	2	1	1	1	1
14	2	2	2	2	1	1	1	1	1	1	1	1	2	2	2	2	2	2	2	2	1	1	1	1	1	1	1	1	2	2	2	2
15	1	1	1	1	2	2	2	2	2	2	2	2	1	1	1	1	2	2	2	2	1	1	1	1	1	1	1	1	2	2	2	2
16	2	2	2	2	1	1	1	1	1	1	1	1	2	2	2	2	1	1	1	1	2	2	2	2	2	2	2	2	1	1	1	1
17	1	1	2	2	1	1	2	2	1	1	2	2	1	1	2	2	1	1	2	2	1	1	2	2	1	1	2	2	1	1	2	2
18	2	2	1	1	2	2	1	1	2	2	1	1	2	2	1	1	2	2	1	1	2	2	1	1	2	2	1	1	2	2	1	1
19	1	1	2	2	1	1	2	2	1	1	2	2	1	1	2	2	2	2	1	1	2	2	1	1	2	2	1	1	2	2	1	1
20	2	2	1	1	2	2	1	1	2	2	1	1	2	2	1	1	1	1	2	2	1	1	2	2	1	1	2	2	1	1	2	2
21	1	1	2	2	1	1	2	2	2	2	1	1	2	2	1	1	1	1	2	2	1	1	2	2	2	2	1	1	2	2	1	1
22	2	2	1	1	2	2	1	1	1	1	2	2	1	1	2	2	2	2	1	1	2	2	1	1	1	1	2	2	1	1	2	2
23	1	1	2	2	1	1	2	2	2	2	1	1	2	2	1	1	2	2	1	1	2	2	1	1	1	1	2	2	1	1	2	2
24	2	2	1	1	2	2	1	1	1	1	2	2	1	1	2	2	1	1	2	2	1	1	2	2	2	2	1	1	2	2	1	1
25	1	1	2	2	2	2	1	1	1	1	2	2	2	2	1	1	1	1	2	2	2	2	1	1	1	1	2	2	2	2	1	1
26	2	2	1	1	1	1	2	2	2	2	1	1	1	1	2	2	2	2	1	1	1	1	2	2	2	2	1	1	1	1	2	2
27	1	1	2	2	2	2	1	1	1	1	2	2	2	2	1	1	2	2	1	1	1	1	2	2	2	2	1	1	1	1	2	2
28	2	2	1	1	1	1	2	2	2	2	1	1	1	1	2	2	1	1	2	2	2	2	1	1	1	1	2	2	2	2	1	1
29	1	1	2	2	2	2	1	1	2	2	1	1	1	1	2	2	1	1	2	2	2	2	1	1	2	2	1	1	1	1	2	2
30	2	2	1	1	1	1	2	2	1	1	2	2	2	2	1	1	2	2	1	1	1	1	2	2	1	1	2	2	2	2	1	1
31	1	1	2	2	2	2	1	1	2	2	1	1	1	1	2	2	2	2	1	1	1	1	2	2	1	1	2	2	2	2	1	1
32	2	2	1	1	1	1	2	2	1	1	2	2	2	2	1	1	1	1	2	2	2	2	1	1	2	2	1	1	1	1	2	2
33	1	2	1	2	1	2	1	2	1	2	1	2	1	2	1	2	1	2	1	2	1	2	1	2	1	2	1	2	1	2	1	2
34	2	1	2	1	2	1	2	1	2	1	2	1	2	1	2	1	2	1	2	1	2	1	2	1	2	1	2	1	2	1	2	1
35	1	2	1	2	1	2	1	2	1	2	1	2	1	2	1	2	2	1	2	1	2	1	2	1	2	1	2	1	2	1	2	1
36	2	1	2	1	2	1	2	1	2	1	2	1	2	1	2	1	1	2	1	2	1	2	1	2	1	2	1	2	1	2	1	2
37	1	2	1	2	1	2	1	2	2	1	2	1	2	1	2	1	1	2	1	2	1	2	1	2	2	1	2	1	2	1	2	1
38	2	1	2	1	2	1	2	1	1	2	1	2	1	2	1	2	2	1	2	1	2	1	2	1	1	2	1	2	1	2	1	2
39	1	2	1	2	1	2	1	2	2	1	2	1	2	1	2	1	2	1	2	1	2	1	2	1	1	2	1	2	1	2	1	2
40	2	1	2	1	2	1	2	1	1	2	1	2	1	2	1	2	1	2	1	2	1	2	1	2	2	1	2	1	2	1	2	1

(Continued)

$$\boxed{L_{64}\ (2^{63})\ \text{(Continued)}}$$

Expt. No.	32	33	34	35	36	37	38	39	40	41	42	43	44	45	46	47	48	49	50	51	52	53	54	55	56	57	58	59	60	61	62	63
41	1	2	1	2	2	1	2	1	1	2	1	2	2	1	2	1	1	2	1	2	2	1	2	1	1	2	1	2	2	1	2	1
42	2	1	2	1	1	2	1	2	2	1	2	1	1	2	1	2	2	1	2	1	1	2	1	2	2	1	2	1	1	2	1	2
43	1	2	1	2	2	1	2	1	1	2	1	2	2	1	2	1	2	1	2	1	1	2	1	2	2	1	2	1	1	2	1	2
44	2	1	2	1	1	2	1	2	2	1	2	1	1	2	1	2	1	2	1	2	2	1	2	1	1	2	1	2	2	1	2	1
45	1	2	1	2	2	1	2	1	1	2	1	1	2	1	2	1	2	1	2	1	2	1	2	1	2	1	1	2	1	2
46	2	1	2	1	1	2	1	2	1	2	1	2	2	1	2	1	2	1	2	1	1	2	1	2	1	2	1	2	2	1	2	1
47	1	2	1	2	2	1	2	1	2	1	2	1	1	2	1	2	2	1	2	1	1	2	1	2	1	2	1	2	2	1	2	1
48	2	1	2	1	1	2	1	2	1	2	1	2	2	1	2	1	1	2	1	2	2	1	2	1	2	1	2	1	1	2	1	2
49	1	2	2	1	1	2	2	1	1	2	2	1	1	2	2	1	1	2	2	1	1	2	2	1	1	2	2	1	1	2	2	1
50	2	1	1	2	2	1	1	2	2	1	1	2	2	1	1	2	2	1	1	2	2	1	1	2	2	1	1	2	2	1	1	2
51	1	2	2	1	1	2	2	1	1	2	2	1	1	2	2	1	1	2	2	1	1	2	2	1	1	2	2	1	1	2	2	1
52	2	1	1	2	2	1	1	2	2	1	1	2	1	2	2	1	1	2	2	1	2	1	1	2	2	1	1	2	2	1	1	2
53	1	2	2	1	1	2	2	1	2	1	1	2	2	1	1	2	2	1	1	2	2	1	1	2	1	2	2	1	1	2	2	1
54	2	1	1	2	2	1	1	2	1	2	2	1	1	2	2	1	2	1	1	2	2	1	1	2	1	2	2	1	1	2	2	1
55	1	2	2	1	1	2	2	1	2	1	1	2	2	1	1	2	2	1	1	2	2	1	1	2	1	2	2	1	1	2	2	1
56	2	1	1	2	2	1	1	2	1	2	2	1	1	2	2	1	1	2	2	1	1	2	2	1	1	2	2	1	1	2	2	1
57	1	2	2	1	2	1	1	2	1	2	2	1	2	1	1	2	1	2	2	1	2	1	1	2	1	2	2	1	2	1	1	2
58	2	1	1	2	1	2	2	1	2	1	1	2	1	2	2	1	2	1	1	2	1	2	2	1	2	1	1	2	1	2	2	1
59	1	2	2	1	2	1	1	2	1	2	2	1	2	1	1	2	1	2	2	1	2	1	1	2	1	2	2	1	2	1	1	2
60	2	1	1	2	1	2	2	1	2	1	1	2	1	2	2	1	2	1	1	2	1	2	2	1	2	1	1	2	1	2	2	1
61	1	2	2	1	2	1	1	2	2	1	1	2	1	2	2	1	1	2	2	1	2	1	1	2	2	1	1	2	1	2	2	1
62	2	1	1	2	1	2	2	1	1	2	2	1	2	1	1	2	2	1	1	2	1	2	2	1	1	2	2	1	2	1	1	2
63	1	2	2	1	2	1	1	2	2	1	1	2	1	2	2	1	1	2	2	1	2	1	1	2	2	1	1	2	1	2	2	1
64	2	1	1	2	1	2	2	1	1	2	2	1	2	1	1	2	1	2	2	1	1	2	2	1	1	2	1	2	2	1	1	2	2	1

L'_{64} (4^{21}) Orthogonal Array

Expt. No.	1	2	3	4	5	6	7	8	9	10	11	12	13	14	15	16	17	18	19	20	21
1	1	1	1	1	1	1	1	1	1	1	1	1	1	1	1	1	1	1	1	1	1
2	1	1	1	1	1	1	2	2	2	2	2	2	2	2	2	2	2	2	2	2	2
3	1	1	1	1	1	1	3	3	3	3	3	3	3	3	3	3	3	3	3	3	3
4	1	1	1	1	1	1	4	4	4	4	4	4	4	4	4	4	4	4	4	4	4
5	1	2	2	2	2	1	1	1	1	2	2	2	2	3	3	3	3	4	4	4	4
6	1	2	2	2	2	2	2	2	2	1	1	1	1	4	4	4	4	3	3	3	3
7	1	2	2	2	2	3	3	3	3	4	4	4	4	1	1	1	1	2	2	2	2
8	1	2	2	2	2	4	4	4	4	3	3	3	3	2	2	2	2	1	1	1	1
9	1	3	3	3	3	1	1	1	1	3	3	3	3	4	4	4	4	2	2	2	2
10	1	3	3	3	3	2	2	2	2	4	4	4	4	3	3	3	3	1	1	1	1
11	1	3	3	3	3	3	3	3	3	1	1	1	1	2	2	2	2	4	4	4	4
12	1	3	3	3	3	4	4	4	4	2	2	2	2	1	1	1	1	3	3	3	3
13	1	4	4	4	4	1	1	1	1	4	4	4	4	2	2	2	2	3	3	3	3
14	1	4	4	4	4	2	2	2	2	3	3	3	3	1	1	1	1	4	4	4	4
15	1	4	4	4	4	3	3	3	3	2	2	2	2	4	4	4	4	1	1	1	1
16	1	4	4	4	4	4	4	4	4	1	1	1	1	3	3	3	3	2	2	2	2
17	2	1	2	3	4	1	2	3	4	1	2	3	4	1	2	3	4	1	2	3	4
18	2	1	2	3	4	2	1	4	3	2	1	4	3	2	1	4	3	2	1	4	3
19	2	1	2	3	4	3	4	1	2	3	4	1	2	3	4	1	2	3	4	1	2
20	2	1	2	3	4	4	3	2	1	4	3	2	1	4	3	2	1	4	3	2	1
21	2	2	1	4	3	1	2	3	4	2	1	4	3	3	4	1	2	4	3	2	1
22	2	2	1	4	3	2	1	4	3	1	2	3	4	4	3	2	1	3	4	1	2
23	2	2	1	4	3	3	4	1	2	4	3	2	1	1	2	3	4	2	1	4	3
24	2	2	1	4	3	4	3	2	1	3	4	1	2	2	1	4	3	1	2	3	4
25	2	3	4	1	2	1	2	3	4	3	4	1	2	4	3	2	1	2	1	4	3
26	2	3	4	1	2	2	1	4	3	4	3	2	1	3	4	1	2	1	2	3	4
27	2	3	4	1	2	3	4	1	2	1	2	3	4	2	1	4	3	4	3	2	1
28	2	3	4	1	2	4	3	2	1	2	1	4	3	1	2	3	4	3	4	1	2
29	2	4	3	2	1	1	2	3	4	4	3	2	1	2	1	4	3	3	4	1	2
30	2	4	3	2	1	2	1	4	3	3	4	1	2	1	2	3	4	4	3	2	1
31	2	4	3	2	1	3	4	1	2	2	1	4	3	4	3	2	1	1	2	3	4
32	2	4	3	2	1	4	3	2	1	1	2	3	4	3	4	1	2	2	1	4	3
33	3	1	3	4	2	1	3	4	2	1	3	4	2	1	3	4	2	1	3	4	2
34	3	1	3	4	2	2	4	3	1	2	4	3	1	2	4	3	1	2	4	3	1
35	3	1	3	4	2	3	1	2	4	3	1	2	4	3	1	2	4	3	1	2	4
36	3	1	3	4	2	4	2	1	3	4	2	1	3	4	2	1	3	4	2	1	3
37	3	2	4	3	1	1	3	4	2	2	4	3	1	3	1	2	4	4	2	1	3
38	3	2	4	3	1	2	4	3	1	1	3	4	2	4	2	1	3	3	1	2	4
39	3	2	4	3	1	3	1	2	4	4	1	1	3	1	3	4	2	2	4	3	1
40	3	2	4	3	1	4	2	1	3	3	2	2	4	2	4	3	1	1	3	4	2

(Continued)

$$L'_{64} \ (4^{21}) \ \text{(Continued)}$$

Expt. No.	Column 1	2	3	4	5	6	7	8	9	10	11	12	13	14	15	16	17	18	19	20	21
41	3	3	1	2	4	1	3	4	2	3	1	2	4	4	2	1	3	2	4	3	1
42	3	3	1	2	4	2	4	3	1	4	2	1	3	3	1	2	4	1	3	4	2
43	3	3	1	2	4	3	1	2	4	1	3	4	2	2	4	3	1	4	2	1	3
44	3	3	1	2	4	4	2	1	3	2	4	3	1	1	3	4	2	3	1	2	4
45	3	4	2	1	3	1	3	4	2	4	2	1	3	2	4	3	1	3	1	2	4
46	3	4	2	1	3	2	4	3	1	3	1	2	4	1	3	4	2	4	2	1	3
47	3	4	2	1	3	3	1	2	4	2	4	3	1	4	2	1	3	1	3	4	2
48	3	4	2	1	3	4	2	1	3	1	3	4	2	3	1	2	4	2	4	3	1
49	4	1	4	2	3	1	4	2	3	1	4	2	3	1	4	2	3	1	4	2	3
50	4	1	4	2	3	2	3	1	4	2	3	1	4	2	3	1	4	2	3	1	4
51	4	1	4	2	3	3	2	4	1	3	2	4	1	3	2	4	1	3	2	4	1
52	4	1	4	2	3	4	1	3	2	4	1	3	2	4	1	3	2	4	1	3	2
53	4	2	3	1	4	1	4	2	3	2	3	1	4	3	2	4	1	4	1	3	2
54	4	2	3	1	4	2	3	1	4	1	4	2	3	4	1	3	2	3	2	4	1
55	4	2	3	1	4	3	2	4	1	4	1	3	2	1	4	2	3	2	3	1	4
56	4	2	3	1	4	4	1	3	2	3	2	4	1	2	3	1	4	1	4	2	3
57	4	3	2	4	1	1	4	2	3	3	2	4	1	4	1	3	2	2	3	1	4
58	4	3	2	4	1	2	3	1	4	4	1	3	2	3	2	4	1	1	4	2	3
59	4	3	2	4	1	3	2	4	1	1	4	2	3	2	3	1	4	4	1	3	2
60	4	3	2	4	1	4	1	3	2	2	3	1	4	1	4	2	3	3	2	4	1
61	4	4	1	3	2	1	4	2	3	4	1	3	2	2	3	1	4	3	2	4	1
62	4	4	1	3	2	2	3	1	4	3	2	4	1	1	4	2	3	4	1	3	2
63	4	4	1	3	2	3	2	4	1	2	3	1	4	4	1	3	2	1	4	2	3
64	4	4	1	3	2	4	1	3	2	1	4	2	3	3	2	4	1	2	3	1	4

L_{81} (3^{40}) Orthogonal Array

Expt. No.	1	2	3	4	5	6	7	8	9	10	11	12	13	14	15	16	17	18	19	20
1	1	1	1	1	1	1	1	1	1	1	1	1	1	1	1	1	1	1	1	1
2	1	1	1	1	1	1	1	1	1	1	1	1	2	2	2	2	2	2	2	2
3	1	1	1	1	1	1	1	1	1	1	1	1	3	3	3	3	3	3	3	3
4	1	1	1	1	2	2	2	2	2	2	2	2	1	1	1	1	1	1	1	1
5	1	1	1	1	2	2	2	2	2	2	2	2	2	2	2	2	2	2	2	2
6	1	1	1	1	2	2	2	2	2	2	2	2	3	3	3	3	3	3	3	3
7	1	1	1	1	3	3	3	3	3	3	3	3	1	1	1	1	1	1	1	1
8	1	1	1	1	3	3	3	3	3	3	3	3	2	2	2	2	2	2	2	2
9	1	1	1	1	3	3	3	3	3	3	3	3	3	3	3	3	3	3	3	3
10	1	2	2	2	1	1	1	2	2	2	3	3	3	1	1	1	2	2	2	3
11	1	2	2	2	1	1	1	2	2	2	3	3	3	2	2	2	3	3	3	1
12	1	2	2	2	1	1	1	2	2	2	3	3	3	3	3	3	1	1	1	2
13	1	2	2	2	2	2	2	3	3	3	1	1	1	1	1	1	2	2	2	3
14	1	2	2	2	2	2	2	3	3	3	1	1	1	2	2	2	3	3	3	1
15	1	2	2	2	2	2	2	3	3	3	1	1	1	3	3	3	1	1	1	2
16	1	2	2	2	3	3	3	1	1	1	2	2	2	1	1	1	2	2	2	3
17	1	2	2	2	3	3	3	1	1	1	2	2	2	2	2	2	3	3	3	1
18	1	2	2	2	3	3	3	1	1	1	2	2	2	3	3	3	1	1	1	2
19	1	3	3	3	1	1	1	3	3	3	2	2	2	1	1	1	3	3	3	2
20	1	3	3	3	1	1	1	3	3	3	2	2	2	2	2	2	1	1	1	3
21	1	3	3	3	1	1	1	3	3	3	2	2	2	3	3	3	2	2	2	1
22	1	3	3	3	2	2	2	1	1	1	3	3	3	1	1	1	3	3	3	2
23	1	3	3	3	2	2	2	1	1	1	3	3	3	2	2	2	1	1	1	3
24	1	3	3	3	2	2	2	1	1	1	3	3	3	3	3	3	2	2	2	1
25	1	3	3	3	3	3	3	2	2	2	1	1	1	1	1	1	3	3	3	2
26	1	3	3	3	3	3	3	2	2	2	1	1	1	2	2	2	1	1	1	3
27	1	3	3	3	3	3	3	2	2	2	1	1	1	3	3	3	2	2	2	1
28	2	1	2	3	1	2	3	1	2	3	1	2	3	1	2	3	1	2	3	1
29	2	1	2	3	1	2	3	1	2	3	1	2	3	2	3	1	2	3	1	2
30	2	1	2	3	1	2	3	1	2	3	1	2	3	3	1	2	3	1	2	3
31	2	1	2	3	2	3	1	2	3	1	2	3	1	1	2	3	1	2	3	1
32	2	1	2	3	2	3	1	2	3	1	2	3	1	2	3	1	2	3	1	2
33	2	1	2	3	2	3	1	2	3	1	2	3	1	3	1	2	3	1	2	3
34	2	1	2	3	3	1	2	3	1	2	3	1	2	1	2	3	1	2	3	1
35	2	1	2	3	3	1	2	3	1	2	3	1	2	2	3	1	2	3	1	2
36	2	1	2	3	3	1	2	3	1	2	3	1	2	3	1	2	3	1	2	3
37	2	2	3	1	1	2	3	2	3	1	3	1	2	1	2	3	2	3	1	3
38	2	2	3	1	1	2	3	2	3	1	3	1	2	2	3	1	3	1	2	1
39	2	2	3	1	1	2	3	2	3	1	3	1	2	3	1	2	1	2	3	2

(Continued)

$L_{81} (3^{40})$ (Continued)

Expt. No.	Column																			
	1	2	3	4	5	6	7	8	9	10	11	12	13	14	15	16	17	18	19	20
40	2	2	3	1	2	3	1	3	1	2	1	2	3	1	2	3	2	3	1	3
41	2	2	3	1	2	3	1	3	1	2	1	2	3	2	3	1	3	1	2	1
42	2	2	3	1	2	3	1	3	1	2	1	2	3	3	1	2	1	2	3	2
43	2	2	3	1	3	1	2	1	2	3	2	3	1	1	2	3	2	3	1	3
44	2	2	3	1	3	1	2	1	2	3	2	3	1	2	3	1	3	1	2	1
45	2	2	3	1	3	1	2	1	2	3	2	3	1	3	1	2	1	2	3	2
46	2	3	1	2	1	2	3	3	1	2	2	3	1	1	2	3	3	1	2	2
47	2	3	1	2	1	2	3	3	1	2	2	3	1	2	3	1	1	2	3	3
48	2	3	1	2	1	2	3	3	1	2	2	3	1	3	1	2	2	3	1	1
49	2	3	1	2	2	3	1	1	2	3	3	1	2	1	2	3	3	1	2	2
50	2	3	1	2	2	3	1	1	2	3	3	1	2	2	3	1	1	2	3	3
51	2	3	1	2	2	3	1	1	2	3	3	1	2	3	1	2	2	3	1	1
52	2	3	1	2	3	1	2	2	3	1	1	2	3	1	2	3	3	1	2	2
53	2	3	1	2	3	1	2	2	3	1	1	2	3	2	3	1	1	2	3	3
54	2	3	1	2	3	1	2	2	3	1	1	2	3	3	1	2	2	3	1	1
55	3	1	3	2	1	3	2	1	3	2	1	3	2	1	3	2	1	3	2	1
56	3	1	3	2	1	3	2	1	3	2	1	3	2	2	1	3	2	1	3	2
57	3	1	3	2	1	3	2	1	3	2	1	3	2	3	2	1	3	2	1	3
58	3	1	3	2	2	1	3	2	1	3	2	1	3	1	3	2	1	3	2	1
59	3	1	3	2	2	1	3	2	1	3	2	1	3	2	1	3	2	1	3	2
60	3	1	3	2	2	1	3	2	1	3	2	1	3	3	2	1	3	2	1	3
61	3	1	3	2	3	2	1	3	2	1	3	2	1	1	3	2	1	3	2	1
62	3	1	3	2	3	2	1	3	2	1	3	2	1	2	1	3	2	1	3	2
63	3	1	3	2	3	2	1	3	2	1	3	2	1	3	2	1	3	2	1	3
64	3	2	1	3	1	3	2	2	1	3	3	2	1	1	3	2	2	1	3	3
65	3	2	1	3	1	3	2	2	1	3	3	2	1	2	1	3	3	2	1	1
66	3	2	1	3	1	3	2	2	1	3	3	2	1	3	2	1	1	3	2	2
67	3	2	1	3	2	1	3	3	2	1	1	3	2	1	3	2	2	1	3	3
68	3	2	1	3	2	1	3	3	2	1	1	3	2	2	1	3	3	2	1	1
69	3	2	1	3	2	1	3	3	2	1	1	3	2	3	2	1	1	3	2	2
70	3	2	1	3	3	2	1	1	3	2	2	1	3	1	3	2	2	1	3	3
71	3	2	1	3	3	2	1	1	3	2	2	1	3	2	1	3	3	2	1	1
72	3	2	1	3	3	2	1	1	3	2	2	1	3	3	2	1	1	3	2	2
73	3	3	2	1	1	3	2	3	2	1	2	1	3	1	3	2	3	2	1	2
74	3	3	2	1	1	3	2	3	2	1	2	1	3	2	1	3	1	3	2	3
75	3	3	2	1	1	3	2	3	2	1	2	1	3	3	2	1	2	1	3	1
76	3	3	2	1	2	1	3	1	3	2	3	2	1	1	3	2	3	2	1	2
77	3	3	2	1	2	1	3	1	3	2	3	2	1	2	1	3	1	3	2	3
78	3	3	2	1	2	1	3	1	3	2	3	2	1	3	2	1	2	1	3	1
79	3	3	2	1	3	2	1	2	1	3	1	3	2	1	3	2	3	2	1	2
80	3	3	2	1	3	2	1	2	1	3	1	3	2	2	1	3	1	3	2	3
81	3	3	2	1	3	2	1	2	1	3	1	3	2	3	2	1	2	1	3	1

(Continued)

$$L_{81} \ (3^{40}) \ \text{(Continued)}$$

Expt. No.	21	22	23	24	25	26	27	28	29	30	31	32	33	34	35	36	37	38	39	40
1	1	1	1	1	1	1	1	1	1	1	1	1	1	1	1	1	1	1	1	1
2	2	2	2	2	2	2	2	2	2	2	2	2	2	2	2	2	2	2	2	2
3	3	3	3	3	3	3	3	3	3	3	3	3	3	3	3	3	3	3	3	3
4	1	1	2	2	2	2	2	2	2	2	2	3	3	3	3	3	3	3	3	3
5	2	2	3	3	3	3	3	3	3	3	3	1	1	1	1	1	1	1	1	1
6	3	3	1	1	1	1	1	1	1	1	1	2	2	2	2	2	2	2	2	2
7	1	1	3	3	3	3	3	3	3	3	3	2	2	2	2	2	2	2	2	2
8	2	2	1	1	1	1	1	1	1	1	1	3	3	3	3	3	3	3	3	3
9	3	3	2	2	2	2	2	2	2	2	2	1	1	1	1	1	1	1	1	1
10	3	3	1	1	1	2	2	2	3	3	3	1	1	1	2	2	2	3	3	3
11	1	1	2	2	2	3	3	3	1	1	1	2	2	2	3	3	3	1	1	1
12	2	2	3	3	3	1	1	1	2	2	2	3	3	3	1	1	1	2	2	2
13	3	3	2	2	2	3	3	3	1	1	1	3	3	3	1	1	1	2	2	2
14	1	1	3	3	3	1	1	1	2	2	2	1	1	1	2	2	2	3	3	3
15	2	2	1	1	1	2	2	2	3	3	3	2	2	2	3	3	3	1	1	1
16	3	3	3	3	3	1	1	1	2	2	2	2	2	2	3	3	3	1	1	1
17	1	1	1	1	1	2	2	2	3	3	3	3	3	3	1	1	1	2	2	2
18	2	2	2	2	2	3	3	3	1	1	1	1	1	1	2	2	2	3	3	3
19	2	2	1	1	1	3	3	3	2	2	2	1	1	1	3	3	3	2	2	2
20	3	3	2	2	2	1	1	1	3	3	3	2	2	2	1	1	1	3	3	3
21	1	1	3	3	3	2	2	2	1	1	1	3	3	3	2	2	2	1	1	1
22	2	2	2	2	2	1	1	1	3	3	3	3	3	3	2	2	2	1	1	1
23	3	3	3	3	3	2	2	2	1	1	1	1	1	1	3	3	3	2	2	2
24	1	1	1	1	1	3	3	3	2	2	2	2	2	2	1	1	1	3	3	3
25	2	2	3	3	3	2	2	2	1	1	1	2	2	2	1	1	1	3	3	3
26	3	3	1	1	1	3	3	3	2	2	2	3	3	3	2	2	2	1	1	1
27	1	1	2	2	2	1	1	1	3	3	3	1	1	1	3	3	3	2	2	2
28	2	3	1	2	3	1	2	3	1	2	3	1	2	3	1	2	3	1	2	3
29	3	1	2	3	1	2	3	1	2	3	1	2	3	1	2	3	1	2	3	1
30	1	2	3	1	2	3	1	2	3	1	2	3	1	2	3	1	2	3	1	2
31	2	3	2	3	1	2	3	1	2	3	1	3	1	2	3	1	2	3	1	2
32	3	1	3	1	2	3	1	2	3	1	2	1	2	3	1	2	3	1	2	3
33	1	2	1	2	3	1	2	3	1	2	3	2	3	1	2	3	1	2	3	1
34	2	3	3	1	2	3	1	2	3	1	2	2	3	1	2	3	1	2	3	1
35	3	1	1	2	3	1	2	3	1	2	3	3	1	2	3	1	2	3	1	2
36	1	2	2	3	1	2	3	1	2	3	1	1	2	3	1	2	3	1	2	3
37	1	2	1	2	3	2	3	1	3	1	2	1	2	3	2	3	1	3	1	2
38	2	3	2	3	1	3	1	2	1	2	3	2	3	1	3	1	2	1	2	3
39	3	1	3	1	2	1	2	3	2	3	1	3	1	2	1	2	3	2	3	1

(Continued)

L_{81} (3^{40}) (Continued)

Expt. No.	Column																			
	21	22	23	24	25	26	27	28	29	30	31	32	33	34	35	36	37	38	39	40
40	1	2	2	3	1	3	1	2	1	2	3	3	1	2	1	2	3	2	3	1
41	2	3	3	1	2	1	2	3	2	3	1	1	2	3	2	3	1	3	1	2
42	3	1	1	2	3	2	3	1	3	1	2	2	3	1	3	1	2	1	2	3
43	1	2	3	1	2	1	2	3	2	3	1	2	3	1	3	1	2	1	2	3
44	2	3	1	2	3	2	3	1	3	1	2	3	1	2	1	2	3	2	3	1
45	3	1	2	3	1	3	1	2	1	2	3	1	2	3	2	3	1	3	1	2
46	3	1	1	2	3	3	1	2	2	3	1	1	2	3	3	1	2	2	3	1
47	1	2	2	3	1	1	2	3	3	1	2	2	3	1	1	2	3	3	1	2
48	2	3	3	1	2	2	3	1	1	2	3	3	1	2	2	3	1	1	2	3
49	3	1	2	3	1	1	2	3	3	1	2	3	1	2	2	3	1	1	2	3
50	1	2	3	1	2	2	3	1	1	2	3	1	2	3	3	1	2	2	3	1
51	2	3	1	2	3	3	1	2	2	3	1	2	3	1	1	2	3	3	1	2
52	3	1	3	1	2	2	3	1	1	2	3	2	3	1	1	2	3	3	1	2
53	1	2	1	2	3	3	1	2	2	3	1	3	1	2	2	3	1	1	2	3
54	2	3	2	3	1	1	2	3	3	1	2	1	2	3	3	1	2	2	3	1
55	3	2	1	3	2	1	3	2	1	3	2	1	3	2	1	3	2	1	3	2
56	1	3	2	1	3	2	1	3	2	1	3	2	1	3	2	1	3	2	1	3
57	2	1	3	2	1	3	2	1	3	2	1	3	2	1	3	2	1	3	2	1
58	3	2	2	1	3	2	1	3	2	1	3	3	2	1	3	2	1	3	2	1
59	1	3	3	2	1	3	2	1	3	2	1	1	3	2	1	3	2	1	3	2
60	2	1	1	3	2	1	3	2	1	3	2	2	1	3	2	1	3	2	1	3
61	3	2	3	2	1	3	2	1	3	2	1	2	1	3	2	1	3	2	1	3
62	1	3	1	3	2	1	3	2	1	3	2	3	2	1	3	2	1	3	2	1
63	2	1	2	1	3	2	1	3	2	1	3	1	3	2	1	3	2	1	3	2
64	2	1	1	3	2	2	1	3	3	2	1	1	3	2	2	1	3	3	2	1
65	3	2	2	1	3	3	2	1	1	3	2	2	1	3	3	2	1	1	3	2
66	1	3	3	2	1	1	3	2	2	1	3	3	2	1	1	3	2	2	1	3
67	2	1	2	1	3	3	2	1	1	3	2	3	2	1	1	3	2	2	1	3
68	3	2	3	2	1	1	3	2	2	1	3	1	3	2	2	1	3	3	2	1
69	1	3	1	3	2	2	1	3	3	2	1	2	1	3	3	2	1	1	3	2
70	2	1	3	2	1	1	3	2	2	1	3	2	1	3	2	1	1	3	2	2
71	3	2	1	3	2	2	1	3	3	2	1	3	2	1	3	2	2	1	3	3
72	1	3	2	1	3	3	2	1	1	3	2	1	3	2	1	3	3	2	1	1
73	1	3	1	3	2	3	2	1	2	1	3	1	3	2	3	2	1	2	1	3
74	2	1	2	1	3	1	3	2	3	2	1	2	1	3	1	3	2	3	2	1
75	3	2	3	2	1	2	1	3	1	3	2	3	2	1	2	1	3	1	3	2
76	1	3	2	1	3	2	3	2	1	3	2	1	3	2	1	2	1	3	1	3
77	2	1	3	2	1	3	1	3	2	1	3	2	1	3	2	3	2	1	2	1
78	3	2	1	3	2	1	2	1	3	2	1	3	2	1	3	1	3	2	3	2
79	1	3	3	2	1	2	1	3	1	3	2	2	1	3	1	3	2	3	2	1
80	2	1	1	3	2	3	2	1	2	1	3	3	2	1	2	1	3	1	3	2
81	3	2	2	1	3	1	3	2	3	2	1	1	3	2	3	2	1	2	1	3

REFERENCES

A1. Addelman, S. "Orthogonal Main Effect Plans for Asymmetrical Factorial Experiments." *Technometrics* (1962) vol. 4: pp. 21-46.

A2. Anderson, B. "Parameter Design Solution for Analog In-Circuit Testing Problems." *Proceedings of IEEE International Communications Conference*, Philadelphia, Pennsylvania (June 1988) pp. 0836-0840.

B1. Box, G. E. P. "Signal to Noise Ratios, Performance Criteria and Transformations." *Technometrics* (February 1988) vol. 30, no. 1, pp. 1-31.

B2. Box, G. E. P. and Draper, N. R. *Evolutionary Operations: A Statistical Method for Process Improvement.* New York: John Wiley and Sons, 1969.

B3. Box, G. E. P., Hunter, W. G., and Hunter, J. S. *Statistics for Experimenters—An Introduction to Design, Data Analysis and Model Building.* New York: John Wiley and Sons, 1978.

B4. Byrne, D. M. and Taguchi, S. "The Taguchi Approach to Parameter Design." *ASQC Transactions of Annual Quality Congress,* Anaheim, CA, May, 1986.

C1. Clausing, D. P. "Taguchi Methods Integrated into the Improved Total Development." *Proceedings of IEEE International Conference on Communications.* Philadelphia, Pennsylvania (June 1988) pp. 0826-0832.

C2. Clausing, D. P. "Design for Latitude." Internal Memorandum, Xerox Corp., 1980.

C3. Cochran, W. G. and Cox, G. M. *Experimental design.* New York: John Wiley and Sons, 1957.

C4. Cohen, L. "Quality Function Development and Application Perspective from Digital Equipment Corporation." *National Productivity Review*, vol. 7, no. 3 (Summer, 1988), pp. 197-208.

C5. Crosby, P. *Quality is Free.* New York: McGraw-Hill Book Co., 1979.

D1. Daniel, C. *Applications of Statistics to Industrial Experimentation.* New York: John Wiley and Sons, 1976.

D2. Deming, W. E. *Quality, Productivity, and Competitive Position.* Cambridge: Massachusetts Institute of Technology, Center for Advanced Engineering Study, 1982.

D3. Diamond, W. J. *Practical Experiment Design for Engineers and Scientists.* Lifetime Learning Publications, 1981.

D4. Draper, N. and Smith, W. *Applied Regression Analysis.*

D5. Duncan, A. J. *Quality Control and Industrial Statistics,* 4th Edition. Homewood, Illinois: Richard D. Irwin, Inc., 1974.

F1. Feigenbaum, A. V. *Total Quality Control,* 3rd Edition. New York: McGraw Hill Book Company, 1983.

G1. Garvin, D. A. "What Does Product Quality Really Mean?" *Sloan Management Review,* Fall 1984, pp. 25-43.

G2. Grant, E. L. *Statistical Quality Control,* 2nd Edition. New York: McGraw Hill Book Co., 1952.

H1. Hauser, J. R. and Clausing, D. "The House of Quality." *Harvard Business Review* (May - June 1988) vol. 66, no. 3, pp. 63-73.

H2. Hicks, C. R. *Fundamental Concepts in the Design of Experiments.* New York: Holt, Rinehart and Winston, 1973.

H3. Hogg, R. V. and Craig, A. T. *Introduction to Mathematical Statistics,* 3rd Edition. New York: Macmillan Publishing Company, 1970.

J1. Jessup, P. "The Value of Continuing Improvement." *Proceedings of the IEEE International Communications Conference,* ICC-85, Chicago, Illinois (June 1985).

J2. John, P. W. M. *Statistical Design and Analysis of Experiments.* New York: Macmillan Publishing Company, 1971.

J3. Juran, J. M. *Quality Control Handbook.* New York: McGraw Hill Book Co., 1979.

K1. Kackar, R. N. "Off-line Quality Control, Parameter Design and the Taguchi Method." *Journal of Quality Technology* (Oct. 1985) vol. 17, no. 4, pp. 176-209.

K2.   Kackar, R. N. "Taguchi's Quality Philosophy: Analysis and Commentary." *Quality Progress* (Dec. 1986) pp. 21-29.

K3.   Katz, L. E. and Phadke, M. S. "Macro-quality with Micro-money." *AT&T Bell Labs Record* (Nov. 1985) pp. 22-28.

K4.   Kempthorne, O. *The Design and Analysis of Experiments.* New York: Robert E. Krieger Publishing Co., 1979.

K5.   Klingler, W. J. and Nazaret, W. A. "Tuning Computer Systems for Maximum Performance: A Statistical Approach." *Computer Science and Statistics: Proceedings of the 18th Symposium of the Interface*, Fort Collins, Colorado (March 1985) pp. 390-396.

L1.   Lee, N. S., Phadke, M. S., and Keny, R. S. "An Expert System for Experimental Design: Automating the Design of Orthogonal Array Experiments." *ASQC Transactions of Annual Quality Congress,* Minneapolis, Minnesota (May 1987) pp. 270-277.

L2.   Leon, R. V., Shoemaker, A. C., and Kackar, R. N. "Performance Measures Independent of Adjustments." *Technometrics* (August 1987) vol. 29, no. 3, pp. 253-265.

L3.   Lin, K. M. and Kackar, R. N. "Wave Soldering Process Optimization by Orthogonal Array Design Method." *Electronic Packing and Production* (Feb. 1985) pp. 108-115.

M1.   Mitchell, J. P. "Reliability of Isolated Clearance Defects on Printed Circuit Boards." *Proceedings of Printed Circuit World Convention IV* (June 1987) Tokyo, Japan, pp. 50.1-50.16.

M2.   Myers, R. H. *Response Surface Methodology.* Blacksburg, Virginia: R.H. Myers, Virginia Polytechnic Institute and State University, 1976.

N1.   Nair, V. N. "Testing in Industrial Experiments with Ordered Categorical Data." *Technometrics* (November, 1986) vol. 28, no. 4, pp. 283-291.

N2.   Nair, V. N. and Pregibon, D. "A Data Analysis Strategy for Quality Engineering Experiments." *AT&T Technical Journal* (May/June 1986) vol. 65, No. 3, pp. 73-84.

P1.   Pao, T. W., Phadke, M. S., and Sherrerd, C. S. "Computer Response Time Optimization Using Orthogonal Array Experiments." *IEEE International Communications Conference.* Chicago, IL (June 23-26, 1985) Conference Record, vol. 2, pp. 890-895.

P2.   Phadke, M. S. "Quality Engineering Using Design of Experiments." *Proceedings of the American Statistical Association, Section on Statistical Education* (August 1982) Cincinnati, OH, pp. 11-20.

P3.   Phadke, M. S. "Design Optimization Case Studies." *AT&T Technical Journal* (March/April 1986) vol. 65, no. 2, pp. 51-68.

P4.   Phadke, M. S. and Dehnad, K. "Optimization of Product and Process Design for Quality and Cost." *Quality and Reliability Engineering International* (April-June 1988) vol. 4, no. 2, pp. 105-112.

P5.   Phadke, M. S., Kackar, R. N., Speeney, D. V., and Grieco, M. J. "Off-Line Quality Control in Integrated Circuit Fabrication Using Experimental Design." *The Bell System Technical Journal,* (May-June 1983) vol. 62, no. 5, pp. 1273-1309.

P6.   Phadke, M. S., Swann, D. W., and Hill, D. A. "Design and Analysis of an Accelerated Life Test Using Orthogonal Arrays." Paper presented at the 1983 Annual Meeting of the American Statistical Association, Toronto, Canada.

P7.   Phadke, M. S. and Taguchi, G. "Selection of Quality Characteristics and S/N Ratios for Robust Design." *Conference Record, GLOBECOM 87 Meeting, IEEE Communications Society.* Tokyo, Japan (November 1987) pp. 1002-1007.

P8.   Plackett, R. L. and Burman, J. P. "The Design of Optimal Multifactorial Experiments." *Biometrika*, vol. 33, pp. 305-325.

P9.   Proceedings of Supplier Symposia on Taguchi Methods, April 1984, November 1984, October 1985, October 1986, October 1987, and October 1988, American Supplier Institute, Inc., 6 Parklane Blvd., Suite 411, Dearborn, MI 48126.

R1.   Raghavarao, D. *Construction of Combinatorial Problems in Design Experiments.* New York: John Wiley and Sons, 1971.

R2.   Rao, C. R. "Factorial Experiments Derivable from Combinatorial Arrangements of Arrays." *Journal of Royal Statistical Society* (1947) Series B, vol. 9, pp. 128-139.

R3.   Rao, C. R. *Linear Statistical Inference and Its Applications,* 2nd Edition. New York: John Wiley and Sons, Inc., 1973.

S1.   Scheffé, H. *Analysis of Variance.* New York: John Wiley and Sons, Inc., 1959.

S2.   Searle, S. R. *Linear Models.* New York: John Wiley and Sons, 1971.

S3.   Seiden, E. "On the Problem of Construction of Orthogonal Arrays." *Annals of Mathematical Statistics* (1954) vol. 25, pp. 151-156.

S4.   Seshadri, V. "Application of the Taguchi Method for Facsimile Performance Characterization on AT&T International Network." *Proceedings of IEEE International Conference on Communications,* Philadelphia, Pennsylvania (June 1988) pp. 0833-0835.

S5.   Sullivan, L. P. "Reducing Variability: A New Approach to Quality." *Quality Progress*, (July 1984) pp. 15-21.

S6.   Sullivan, L. P. "Quality Function Deployment." *Quality Progress* (June 1986) pp. 39-50.

T1. Taguchi, G. *Jikken Keikakuho,* 3rd Edition. Tokyo, Japan: Maruzen, vol. 1 and 2, 1977 and 1978 (in Japanese). English translation: Taguchi, G. *System of Experimental Design,* Edited by Don Clausing. New York: UNIPUB/Kraus International Publications, vol. 1 and 2, 1987.

T2. Taguchi, G. "Off-line and On-Line Quality Control System." *International Conference on Quality Control.* Tokyo, Japan, 1978.

T3. Taguchi, G. *On-line Quality Control During Production.* Tokyo, Japan: Japanese Standards Association, 1981. (Available from the American Supplier Institute, Inc., Dearborn, MI).

T4. Taguchi, G. *Introduction to Quality Engineering.* Asian Productivity Organization, 1986. (Distributed by American Supplier Institute, Inc., Dearborn, MI).

T5. Taguchi, G. and Konishi, S. *Orthogonal Arrays and Linear Graphs.* Dearborn, MI: ASI Press, 1987.

T6. Taguchi, G. and Phadke, M. S. "Quality Engineering through Design Optimization." *Conference Record, GLOBECOM 84 Meeting, IEEE Communications Society.* Atlanta, GA (November 1984) pp. 1106-1113.

T7. Taguchi G. and Wu, Yu-In. *Introduction to Off-Line Quality Control.* Central Japan Quality Control Association. Meieki Nakamura-Ku Magaya, Japan, 1979. (Available from American Supplier Institute, Inc., Dearborn, MI).

T8. *The Asahi*, Japanese language newspaper, April 15, 1979. Reported by Genichi Taguchi during lectures at AT&T Bell Laboratories in 1980.

T9. Tomishima, A. "Tolerance Design by Performance Analysis Design—An Example of Temperature Control Device." *Reliability Design Case Studies for New Product Development,* Edited by G. Taguchi, Japanese Standards Assoc., 1984, pp. 213-220 (in Japanese).

Y1. Yokoyama, Y. and Taguchi, G. *Business Data Analysis: Experimental Regression Analysis.* Tokyo: Maruzen, 1975.

INDEX